COVER: *This picture, painted by western artist and gold prospector Albertis Browere in 1854, shows two ways in which a group of prospectors are extracting placer gold—loose particles of gold—from gravel. The man seated at right is working a cradle (a gravel-washing machine); the man in the center, probably the artist, is panning gold.*

PAGES 2 AND 3: *Before stores were set up in the California mining fields, the miners had to depend on supply wagons like this one to bring in the tools and the food they needed. These wagons also hauled the gold dust from mines to refineries.*

TITLE PAGE: *The lone gold prospector—his tools, pan, and provisions loaded on the back of his sure-footed mule—was a very familiar figure in the California gold fields before big mining companies were formed and began large-scale operations to find gold.*

PAGES 154 AND 155: *San Francisco had become a boom town of 25,000 people in 1850 when this picture was painted. Emigration into the port of San Francisco had increased when California was claimed as an American territory in 1846; but it became a human tidal wave after 1848 when people learned of the discovery of gold at Sutter's Mill.*

THE CALIFORNIA GOLD RUSH

ILLUSTRATED WITH PAINTINGS, PRINTS, DRAWINGS, MAPS, AND PHOTOGRAPHS OF THE PERIOD

THE CALIFORNIA
GOLD RUSH

BY THE EDITORS OF
AMERICAN HERITAGE

NARRATIVE BY
RALPH K. ANDRIST

IN CONSULTATION WITH
ARCHIBALD HANNA

PUBLISHED BY
TROLL ASSOCIATES
Mahwah, N.J.

Foreword

From the very beginning down to the final closing of the frontier, the westward movement of the pioneers was a part of American life. Year by year the tide crept westward. Sometimes it was slowed by a mountain range or temporarily checked by hostile Indians. Yet it never halted until the whole land was occupied and there was no longer a zone that could be called a frontier between the settlements and the wilderness.

The California gold rush was a part of this great westward movement, but it had certain distinctive features that set it apart from the usual pattern of frontier settlement. One of the most striking of these was the speed with which it was accomplished. On the old frontier it was usually a matter of many years before a new settlement amounted to more than a handful of cabins in the partly cleared forest. But in the first two years after the discovery of gold in California almost one hundred thousand people poured into the new territory—more than the number of people living in the state of Delaware at that time. No new territory had ever been settled so quickly.

Startling, too, was the fact that almost all of these new settlers were men. On the old frontier the social unit was the family. Father, mother, and children moved together to the new lands. But in California, during the first year at least, women were so unusual that miners were known to walk miles to another camp for the sight of one. Many of the forty-niners were young and unmarried, and those who were married had usually left their wives at home. Unlike the average American pioneer, most of the forty-niners had no intention of remaining in California longer than the few months they expected they would need in order to make their "piles." They came not to settle, but to dig enough gold to pay off the mortgage on the farm at home, or to buy a new farm, or to start out in business for themselves. A few were able to do this, though many returned home no richer than when they had come. Yet many more sent for their families and remained in California.

The final distinctive feature of the gold rush lay in the great distance that had to be traveled to reach the new territory. Always before, the frontier had been merely the outer fringe of the settled area, with no real gap between; now it had moved almost fifteen hundred miles west at a single bound, leaving the plains and the Rockies to be settled later.

In the years that followed there were to be other gold rushes; to Pike's Peak in 1859, to Montana in the sixties and to the Klondike in the nineties. But the California gold rush was the first and greatest of them all, and set the basic pattern for the miner's frontier.

The factual accounts and contemporary illustrations brought together in this book should give young readers a good understanding of the decisive manner in which the forty-niners spurred the growth of the West, while also conveying much of the turbulence and enthusiasm of that adventurous era.

ARCHIBALD HANNA

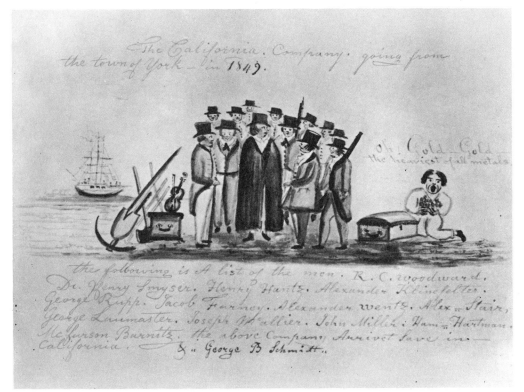

The men in the drawing above are members of the California Company, a group of adventurers who set out from their homes in York, Pennsylvania, in 1849—having pooled their money to pay for tools, supplies, and transportation by ship—to hunt for gold in California. A member of the expedition, Lewis Miller, made this drawing in his journal.

Contents

Sutter's Mill

The news of the discovery of gold in California in 1848 signaled the start of the first gold rush in the history of the world. Within the year 1849 alone, the new American territory's population swelled fivefold as 80,000 men hurried there in the hope of claiming a share of the golden riches. By 1852, the state held nearly a quarter of a million people, and each year the mines continued to produce a treasure worth fifty million dollars; by 1865, three quarters of a billion dollars in gold had been clawed from the hills and stream beds of California.

In other countries and in earlier times, mines that produced precious metals had become the property of kings and princes. But things were done more democratically in Califor-

Telegraph Hill in San Francisco had a tent city on its slopes in 1849 when this painting was made. Men from all over the world landed here, bought supplies, and then went into the hills to look for gold.

nia, and the gold, like everything else on the frontier, belonged to the men who took it.

Probably the best way to begin the story of the gold rush is to tell of the man whose employee first discovered gold and who, in the end, lost everything he had because of it.

His name was Johann August Suter, and although born in Germany in 1803, he was proud to recall that his parents had originally come from Switzerland, one of Europe's few democracies at that time. Things were not going well for Suter in Europe, and in 1834, when he was thirty-one years old and had a wife and four children to support, he was so deeply in debt that he decided to make a new start in America. After putting his family in the care of his brother and promising to send for them when he was able, he left home.

In America, one of the first things he did was to change his name to John Augustus Sutter because he wished to be as American as possible. He headed at once for Missouri, where he soon became a trader on the Santa Fe Trail. In Santa Fe he heard accounts of the Mexican province of Upper California to the west, with its mild climate, its rich soil, and all the land an ambitious man could want. He decided then that he would be a rancher in California, and after three years he gave up being a trader and set out, in 1838, with eight other men, for the Far West to do something about making his dreams come true.

The party followed the Oregon Trail as far west as Fort Vancouver (near the present-day city of Portland, Oregon), but there was no ship there to take Sutter to California. However, the Hudson's Bay Company did have a ship about to sail for the Sandwich Islands—as the Hawaiian Islands were then called—with a cargo of furs. When it set sail Sutter was aboard.

At Honolulu he still could not find a vessel sailing in the direction he wanted to go, but an English ship, the *Clementine,* was available for charter, and Sutter proposed to hire her, load her with trade goods, and go to Sitka, Alaska, where the Russians were eager to buy goods. When the *Clementine* sailed, Sutter had with him ten Kanakas (natives of the Sandwich Islands)

who had agreed to work for him in California for three years. Also aboard were certain articles he had purchased for purposes that looked beyond the trip to Alaska. Among them were three small brass cannon to be used to defend the great ranch that he planned to own in California.

When the trading in Sitka was successfully concluded, the *Clementine* headed south. After a stormy trip, the ship sailed through the Golden Gate on July 1, 1839, and anchored off the little village of Yerba Buena, which was soon to change its name to San Francisco. The military commander refused to let the *Clementine* remain there without an official entry permit, which could only be obtained at the capital, Monterey, down the coast. Sutter put to sea again.

On July 5, 1839, at Monterey, Sutter met the governor, Juan Bautista Alvarado; they were soon on very friendly terms, and Alvarado became interested in what his visitor had to say. Sutter talked about settling east of San Francisco Bay, in a part of California where no one but Indians lived. The land was worthless in the governor's opinion, and besides, it might be a good thing to have a friendly settler there because Russia, Great Britain, and the United States were all showing interest in the same general region.

In 1838 when John Sutter landed at Honolulu in the Hawaiian Islands—in his attempt to get to California by sea—he might well have seen Kinau (center), the wife of the kingdom's high chief, surrounded by her courtiers.

And so John Sutter left Monterey with permission to select a tract of land for settlement. (Nearly two years later, in June, 1841, Alvarado would make Sutter a Mexican citizen, and would grant him about 49,000 acres of land.)

Back at Yerba Buena, Sutter hired two small sloops to carry his supplies, and a four-oared launch for use in shallower water where the sloops might not be able to go. He ordered tools, farming implements, and other equipment, and then set out once more. But he did not head back to the open sea this time; instead, he sailed north and east across San Francisco Bay and into an inlet called Suisun Bay, and entered the mouth of the Sacramento River.

John Sutter took his time. He could choose any location he liked in the entire Sacramento Valley, and he wanted to be sure he chose wisely. For eight days his expedition moved upstream,

In 1846, under the Bear Flag at top, American settlers in California declared themselves independent of Mexican rule. James Marshall (center), who fought in the Bear Flag Revolt, changed the history of California even further by finding gold at Sutter's Mill on January 24, 1848. The mill was owned by John Sutter (above left), the ruler of New Helvetia, a tract of land granted by Juan Bautista Alvarado (above right), the Mexican governor of California, in 1841.

stopping often to leave the boats and explore the land beyond the tree-bordered banks of the river. Once a party of Indians showered them with arrows, but no one was hurt. On the eighth day they came to the place where the American River joins the Sacramento. They had gone only a couple of miles up the American River when Sutter stopped again. He noted the rich, rolling land, dotted with occasional clumps of trees; he observed that a certain small knoll would make an excellent site for a fort. There the boats were unloaded and a camp set up.

During the next few years, Sutter prospered. On the knoll he had noticed that first day, he built his fort, a structure of adobe clay with walls eighteen feet high and three feet thick, with bastioned corners on which the three cannon he had brought from the Sandwich Islands were mounted. On September 4, 1841, the Russian schooner *Constantine* arrived at Sutter's Fort with the governor of the Russian fur-trading posts in California aboard. The Russians wished to sell John Sutter their lands on the coast of California. Their holdings included Fort Ross—eighty miles north of San Francisco—and a trading post at Bodega Bay, which was fifty miles north of San Francisco. Fort Ross, which the Russians had founded in 1812 as a supply base for their northern fur-trading bases in Alaska, was particularly valuable because the property included several buildings, herds of livestock, and farm equipment. By 1841 the Rus-

Misión of St. Louis (the Bishop) about the middle of the old territory.
s. Lui Obispo.

Dec. 27.
Mex.

sian fur trade in Alaskan waters had dwindled considerably, and the cost of maintaining supply bases in California no longer seemed justified. The Russians were willing to sell everything they held in California to Sutter for $30,000. They asked for a down payment of only $2,000, and Sutter agreed to buy. From the fall of 1841 to the spring of 1842 Sutter had the contents of Fort Ross and even most of the buildings themselves dismantled and shipped to his fort. Soon he had blacksmith and carpentry shops, a tannery and a gristmill, a distillery and a blanket-weaving shop in operation. His herds of cattle increased until he had 13,000 head, and large acreages were planted with wheat and other grain. There was a ten-acre orchard with apple, peach, olive, almond, pear, and fig trees, and even two acres of Castille roses, grown from root cuttings that the Mexican mission priests had given him.

The mission of San Luis Obispo de Tolosa, shown above in 1850, was founded by the Spanish in 1772 as one of a series of missions in California intended to Christianize the Indians and to spread Spanish culture.

Sutter called his domain New Helvetia, in honor of Switzerland, whose Latin name is Helvetia. It was almost a little kingdom in its own right, able, if need be, to protect itself against Indians or even against Mexican forces. Sutter's Fort was directly on the overland route from the United States to California. It became an important way station for an increasing number of American emigrants who were beginning to come into northern California even though the Mexican authorities did nothing to encourage them. Sutter, although now a Mexican citizen, helped the Americans by allowing them to use his fort as a kind of headquarters.

As more Americans settled in California their desire for independence

Sutter's Mill (above), a sawmill located on the banks of the American River at Coloma, about forty miles from Sutter's Fort (below), was begun by James Marshall, an employee of Sutter's, in 1847. It was on January 24, 1848, that Marshall found, in the tailrace of the still unfinished sawmill, the first traces of the gold that was to change the history of California and the nation. On January 28, he carried the news to John Sutter.

from Mexico grew steadily stronger. In 1846 two events occurred which changed the history of California: on May 13 war broke out between the United States and Mexico, and on June 14 a group of American settlers at Sonoma, California, declared themselves independent of Mexico and citizens of the California Republic; its symbol was the Bear Flag, which bore a single star and a picture of a grizzly bear. By January 13, 1847, Mexican forces in California had been decisively defeated by United States troops under Colonel Stephen Watts Kearny, and a year later, when the Treaty of Guadalupe Hidalgo ending the Mexican War was signed on February 2, 1848, Mexico ceded California to the United States.

But just the month before that treaty was signed, something else had happened that would bring an end to all the dreams of John Augustus Sutter.

In the summer of 1847, Sutter had decided to build a sawmill farther up the American River, on its south fork, about forty miles northeast of the fort, in the foothills of the Sierra Nevada Mountains. Timber was the one thing his enormous holdings did not supply in quantity, but good trees were plentiful in the mountains. He made an agreement with James Wilson Marshall, a carpenter who worked for him, by which Marshall would build and operate the sawmill in return for a share of the lumber.

Marshall began construction with a crew of workmen at a place called Coloma. The months passed, and the mill rose without anything unusual happening until after the first of the year—the date, January 24, 1848, is an important one in American history. The night before, Marshall had turned the water from the millpond into the tailrace (the ditch that led the water away from the mill wheel) so that the channel would be washed free of debris and loose dirt. In the morning, when he came to look at the results, his eye was caught by a yellow gleam at the bottom of the clear water. He picked out a small bit of yellow metal, and found there were others scattered about.

Marshall put the metal to the simplest test he knew: he pounded it with a rock. It was soft and flattened easily like gold; fool's gold, which is so often mistaken for the real thing, is brittle and breaks into bits. James Marshall

The historic gold nugget that James Marshall first took from the tailrace of Sutter's Mill is seen below in an enlarged photograph. It is actually about the size of a dime and weighs just over a quarter of an ounce.

was not a calm man to begin with, and the more he thought about his discovery, the more excited he became. On January 28 at Sutter's Fort he burst into the office of his employer.

"He was soaked to the skin and dripping water," Sutter wrote in his diary. "He told me he had something of the utmost importance to tell me, that he wanted to speak to me in private, and begged me to take him to some isolated place where no one could possibly overhear us. We went up to the next floor, and, although there was no one in the house except the bookkeeper, he insisted so strongly that we locked ourselves in a room."

Sutter seems to have wondered a little about the sanity of his visitor but did as the latter asked. However, he had to leave the room again, and forgot to lock the door when he came back. Marshall had just taken a piece of cloth from his pocket and was unwrapping it when the bookkeeper happened to walk into the room to ask Sutter a question.

Sutter's diary continues: " ' My God, didn't I tell you to lock the door,' cried Marshall. He was in a terrible state of excitement and I had all the trouble in the world to quiet him and to convince him that the bookkeeper had come on his own business and not to spy on us. This time we bolted the door and even pushed a wardrobe against it."

In 1852 James Marshall posed for the photograph at left, standing in front of Sutter's Mill where gold was first found in 1848. Marshall's discovery was to change completely the future of Sutter's Fort (above), the heart of New Helvetia, which was still a thriving and prosperous place in 1849 when this sketch was made.

Marshall then showed Sutter what he had discovered at the mill. Sutter at first glance thought that Marshall had indeed discovered gold. He records in his diary his words to Marshall: "Well, it looks like gold. Let us test it."

Sutter had an encyclopedia and turned to the section on gold. The two men put the samples to all the tests they could. They pounded it, they weighed it in water, and they tested it with nitric acid to see whether it resisted corrosion. It was, beyond all doubt, gold. Now it was not only Marshall but Sutter as well who was disturbed. The latter sensed that the discovery would mean trouble for him. He loved his land, and he wanted

nothing more out of life than to manage his acres and watch his herds increase. Gold could ruin everything.

Sutter went up to the sawmill to see for himself, possibly hoping that the gold was scarce and that Marshall had just happened to come on a few grains. But it was there, and in quantity. He sighed, for he knew that soon other men would be coming up to look for gold. He could only ask the workmen at the sawmill not to tell anybody about the discovery for at least another six weeks. That would give him a chance to get his spring planting done before his men started to leave to look for gold. After that—who could know what would happen?

Gold! Gold! Gold!

Strangely enough, the workmen at the sawmill did not become immediately excited by the discovery of gold. Through February they continued to work on the sawmill as usual and only looked for gold on Sunday, their day off. After mid-March, when the mill was finished, some of them began to dig full time. Then, just as Sutter had feared, his men began to leave the fort, first by ones and twos and then in a steady stream. Work slowed down, but luckily most of the Indians there had little interest in gold and remained to take care of the fields and herds.

The news quickly reached San Francisco, since there was regular traffic between that city and Sutter's Fort. A small item appeared in the newspaper in mid-March—but on the same day there was a much larger story about the finding of copper deposits. Nobody had yet realized the extent and richness of the gold fields.

One man, however, went out of his way to see that people became excited about the discovery of gold. He was an elder in the Mormon Church and his name was Sam Brannan. He had built a fine store at Sutterville, a settlement on the Sacramento River a few miles below Sutter's Fort. Sam was a businessman and planned to let others dig while he got his portion of the gold by trading with the miners. Men would have to pass Brannan's store on their way to the gold fields, so the more miners he interested in the diggings, the more he would find stopping in at his store for supplies.

Accordingly, Sam Brannan set out one day in May in Sutter's boat headed for San Francisco. With him he carried a bottle of gold dust, and when he reached the city, he walked up and down the streets, showing the gold to everyone he met and shouting, "Gold! Gold from the American River!" By the end of the day, many people in San Francisco were more interested in the discovery at Sutter's Mill than they had been before Brannan's arrival.

After Brannan's visit, a large group of excited citizens boarded a fleet of launches and left for the gold fields. Others started overland. A week later, a boat from the fort docked in San Francisco. Aboard it were large amounts of gold. Those who had earlier resisted Sam Brannan's efforts now gave in completely to the gold fever. Almost everyone who could walk left for the diggings: doctors, lawyers, clergymen, schoolteachers, blacksmiths, tailors, the town's first and second al-

When John Sutter, a major general in the California state militia, posed for this portrait in 1855, his extensive land holdings and his great personal fortune had already been lost.

caldes, or magistrates, and even the sheriff. Fewer than one hundred people were left in San Francisco.

Sam Brannan now had all the business he could handle at his store, and was making more money than most of the miners ever would. But things were not going well for John Sutter. He had always been hospitable and gracious to visitors at the fort, and he tried to treat the passing miners as guests, feeding those who stopped, and lending horses. But with his workmen leaving and the miners coming in increasing numbers, it was becoming impossible to continue helping so many strangers. His diary entries, which had once recorded all the busy events of his great ranches, became shorter and shorter. On May 23 he wrote: "Hosts arriving by water and land for the Mts. Fine day." Two days later he wrote only six words: "Great hosts continue to the Mts." It was the last thing he wrote in the diary. The old, pleasant days had come to an end.

Sutter, immediately after learning that there was gold at his mill in the Coloma Valley, began to worry about prospectors overrunning his property there. He signed a treaty with the Indians of the valley, leasing the land from them for three years. Then, early in March, 1848, he sent one of his men, Charles Bennett, to the new American military governor at Monterey, Richard Mason, in the hope that the United States would confirm his claim to the land around the sawmill. Mason refused to confirm Sutter's claim because

word had not yet reached California that the treaty ending the Mexican War had been signed nor that Mexico had legally ceded California to the United States.

Sutter tried mining in the summer of 1848 with a large crew of Indians and Kanakas. But he was not made to be a miner. Whatever gold he took from the river gravel went to pay his debts, the greater part of which were still owed to the Russians for his purchase of Fort Ross.

With the coming of winter and bad weather, many of Sutter's men came back to work for him, but they were off again as soon as another spring arrived. By April of 1849 only a wagonmaker and a blacksmith remained with him. Then they too were gone, and all he had left were some of his Indians. The fort was almost deserted. Shops and gristmill were silent, and hides rotted in the tanning vats.

Sutter's son, John, Jr., had joined him in late August of 1848, the first member of his family Sutter had seen since he left them fourteen years earlier. It was good to have his son to help him, but there was far more to be done than two men and a crew of Indians could ever hope to do.

Most of the gold seekers who came

This map (right) shows the tributary rivers flowing down through the foothills of the Sierra Nevada Mountains, on the east, into the San Joaquin and Sacramento rivers in California's central valley. It was along these tributaries, where gold strikes were first made, that many small mining towns arose.

THE
CALIFORNIA
GOLD FIELDS

SHOWING SOME OF THE MORE
IMPORTANT MINING CAMPS,
CIRCA 1850

Lake
Almanor

W. Branch

E. Branch

N. Fork

Rich
Bar

Deer Creek

M. Fork

Whiskey Flat

S. Fork

Canyon Creek

Poker
Flat

Dutch Flat

Helltown

Goodyear's
Bar

Bidwell's Bar

Downieville

Cut Eye Foster's Bar

N. Yuba R.

Kanaka Flat

FEATHER R.

Virginia City
(Comstock Lode)

M. Yuba R.

Lousy Level

S. Yuba R.

Washington

*Washoe
Lake*

Rough and Ready

YUBA R.

Nevada City

You Bet

N. Fork

Lake
Tahoe

SACRAMENTO R.

Marysville

Grass
Valley

Iowa Hill

Carson
City

Plumas
City

BEAR R.

Illinoistown

Wisconsin Hill

SIERRA NEVADA

Murderers Bar

M. Fork

Coloma (Sutter's Mill)

NEVADA
CALIFORNIA

Auburn

AMERICAN R. S. FORK

Placerville (Hangtown)

Diamond Springs

Missouri
Flat

Grizzly Flats

Sacramento

Sutter's
Fort

Fair Play

COSUMNES R.

Volcano

N. Fork

M. Fork

Jackson

S. Fork

Rich Gulch

N. Fork

Sonoma

Mokelumne
Hill

Dogtown

M. Fork

San Pablo
Bay

Suisun
Bay

MOKELUMNE R.

CALAVERAS R.

Angels Camp

S. Fork

Shaw's Flat

Sonora

*Golden
Gate*

San Francisco

Rawhide

San
Francisco
Bay

Chinese
Camp

STANISLAUS R.

TUOLUMNE R.

SAN JOAQUIN R.

PACIFIC
OCEAN

MERCED R.

Hornitos

Mormon
Bar

COAST RANGES

MARIPOSA R.

FRESNO R.

N

Scale
.0   5   10   15 Miles

Monterey

COAST RANGES

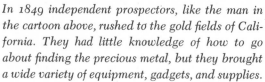

In 1849 independent prospectors, like the man in the cartoon above, rushed to the gold fields of California. They had little knowledge of how to go about finding the precious metal, but they brought a wide variety of equipment, gadgets, and supplies.

The crowded scene above—showing elegantly dressed gentlemen, bearded prospectors, and even Indians—is only a slight exaggeration, for people of every sort came to California, eager to pan for gold.

The cartoon at right, titled California Gold Hunter Meeting a Settler, ridicules the greeting many miners received both from California mountain jaguars—and from permanent residents of California.

Europeans were fascinated by tales of the gold rush and of greedy miners who defended their claims to the death. This swashbuckling illustration (left) is from a French adventure novel of the gold rush.

in 1848 had been from California, knew of Sutter, and respected him. But by 1849 men were arriving from everywhere. This new wave of miners stole Sutter's cattle or butchered them in the fields. They turned their horses and oxen out to graze in his grain fields. They stole without mercy from the fort, even taking three large millstones, probably to be used for crushing gold-bearing rock. A new town, Sacramento City, was founded in 1848 where the American River joins the Sacramento, only a couple of miles from the fort. The city boomed, and while a few of the settlers bought land from Sutter, most of them just took it —and then sold it to others.

John Sutter did what he could to make something from the changed conditions. His boats did a thriving business carrying passengers and freight between San Francisco and Sacramento City, and he rented out the empty stables and shops in the fort to merchants. He also set up stores at Sutterville and at Coloma, where thousands of miners were digging near the site of the sawmill. But he had to hire men to run his stores and haul his goods, and most of them cheated him. Soon, too, mushrooming Sacramento City took leadership away from Sutter's Fort and Sutterville. The traders gave up the space they had rented, and that source of income disappeared too. But shrewd Sam Brannan survived the change in good shape; he abandoned his store in Sutterville and moved to Sacramento City, where he is reported to have made a profit of $160,000 in one year from a hotel he ran.

Sutter leased the fort and moved away to live on a farm he owned on the Feather River. His family finally joined him there in 1851, when he had very little left to show them for his years of work. The titles to his land were confused because his grants had come from the Mexican government. Years later, when the United States Supreme Court upheld his right to at least part of the land, there was nothing left for him after he had paid off obligations to others. He finally received a pension of $3,000 a year from the state of California but had no luck in his attempts to obtain justice from the United States government. He moved to a small town in Pennsylvania called Lititz, and died in Washington in 1880 on one of his many trips to ask Congress for help. To John Augustus Sutter, the gold rush brought nothing but loss and heartbreak.

But in 1848, all this was a long way in the future, and California was interested in little else besides the spreading hunger for gold. In San Francisco, Mr. Buckelew, publisher of the newspaper, *The Californian*, wrote an editorial which appeared on May 29: "The majority of our subscribers and many of our advertisers have closed their doors and places of business and left town. . . . The whole country, from San Francisco to Los Angeles, and from the seashore to the Sierra Nevada, resounds to the sordid cry of Gold! Gold! GOLD! while the field is half

planted, the house half built, and everything neglected but the manufacture of shovels and pickaxes." That was the last issue of *The Californian.* Next day, Mr. Buckelew himself left for the mountains to dig for gold.

As the fever spread, scenes in Sonoma, Monterey, Los Angeles, San Diego, and the other principal towns of California were much like those in San Francisco. Walter Colton, the American alcalde at Monterey, wrote in his diary on May 29: "Our town was startled out of its quiet dreams today by the announcement that gold had been discovered on the American Fork. The men wondered and talked, and the women too; but neither believed." So Colton sent a man up to the American River to learn the truth. On June 20 the messenger returned (which was somewhat unusual as most men would not have come back at all), and the whole town rushed to hear his story. Mr. Colton wrote: "As he drew forth the yellow lumps from his pockets and passed them around among the eager crowd the doubts, which had lingered till now, fled. . . . The excitement produced was intense; and many were soon busy in their hasty preparations for a departure to the mines . . . the blacksmith dropped

his hammer . . . the farmer his sickle, the baker his loaf, and the tapster his bottle. . . . I have only [the] women left, and a gang of prisoners; with here and there a soldier, who will give his captain the slip at the first chance."

The news also traveled north, into Oregon, and settlers there dropped everything to head for the gold fields.

One of the first foreign lands to be affected was the Sandwich Islands, and both whites and Kanakas crowded every ship bound for the Golden Gate. East of California, the news of gold soon reached the Mormon settlement

OVERLEAF: *This picture of Coloma Valley — on the South Fork of the American River in the early 1850's — shows Sutter's Mill at the river's edge (center) and the new mining town of Coloma (left) on both sides of the river.*

of Salt Lake City; young Mormons returning home from California brought impressive amounts of gold and even more impressive stories, and they fanned the fever to a blaze. Brigham Young, the Mormon leader, appealed to his young men to stay, but he did not have much success in keeping them with him in Utah.

During that first year, 1848, prospectors found most of the general areas where there was pay dirt (although those who came later found that many of the original diggings still yielded immense quantities of gold). They found rich deposits along the branches of the Sacramento: the American River, the Feather River, the Yuba. A Captain Charles Weber left the Sacramento Valley and searched along the tributaries of the San Joaquin to the south; the fields there, along the headwaters of the Stanislaus, the Tuolumne, the Consumnes and Mokelumne rivers, were as rich as those of the Sacramento feeder streams. And far to the north, men from Oregon on their way to the gold fields looked in the wild canyons of California's wilderness stream named the Trinity River, and found gold in quantity there.

Not everyone found gold, even that first year when there was not too much competition for the choice spots. Luck played its part in prospecting, and it was not always the first man there who struck it rich. One of those who failed was Claude Chana, one of the Sutter workmen who had headed for the sawmill at Coloma when the excitement

began. From there he moved to the north, looking for gold in the Sierra Nevada, along the deep ravines where tributaries of the North Fork of the American River flowed. He came across one likely-looking place, but like many another miner, he soon moved on looking for something better. Other prospectors came along and found that the deposit Chana had abandoned was very rich; during those summer and fall months of 1848, hard-working miners there were able to take out $800 to $1,500 in a single day. At Chana's former diggings a large settlement grew up and was named Auburn, and just about the time the gold began to run thin and some of the miners moved on, the biggest strike yet was made—completely by accident. A man named Jenkins was working the gravel at a side canyon named Missouri Gulch, but having a hard time of it because of a shortage of water. So he dug a ditch to bring water from the stream to his claim. If the ditch had worked perfectly, Jenkins might have taken out no more than a modest amount of gold, but one section of it caved in by breaking into a hole, probably a ground squirrel burrow. The dirt was washed away to reveal a fabulously rich deposit, from which Jenkins took out $40,000 in one month.

Along the South Fork of the American River, where it all began at the Coloma sawmill, there were two thousand men by June 1, spread out along the river for thirty miles. By the end of July, there were four thousand. The

sawmill itself disappeared, torn apart to get lumber to make gold-washing equipment. A busy town sprang up which in its heyday would contain thirteen hotels while ten thousand men dug in the vicinity. The gold is long gone, but Coloma still remains, commemorated by a statue of James Marshall, pointing to the spot where he found the first flakes of gold.

And what of Marshall, the man who found the first gold and rushed down to Sutter's Fort in such high excitement? He suffered a sadder fate even than John Sutter. James Marshall joined in the search for gold but was never lucky enough to become

These prospectors, photographed in 1852, are using a new device called a long tom (see page 76) to wash placer gold from the gravel of a stream bed at Auburn, California.

wealthy. A very stubborn man, he continued to look while his mind slowly gave way and he came to believe that he was the rightful owner of all the gold in California. As he searched, he became so quarrelsome and short-tempered that other miners would not put up with him and drove him out of one camp after another. He returned to live in a little cabin near the site of his discovery and died there in 1885, long after the gold rush, in complete poverty.

31

The steamer Hartford *(above), her decks crowded with easterners going to seek their fortunes in California, is shown leaving New York harbor in February, 1849. The* Hartford's *owners were pioneers in the attempt to establish a regular New York-to-San Francisco mail and passenger run.*

Forty-niners on the High Seas

On February 28, 1849, the steamship *California* entered the Golden Gate and received a thunderous welcome from the guns of the Pacific Squadron of the U.S. Navy at anchor in San Francisco Bay. Commodore Jones, commanding the squadron, had ordered the salute in honor of the *California*. She was the first steamship to complete the trip around Cape Horn and to arrive at San Francisco. The men who lined her rail and cheered as the smoke from the cannon drifted across the water were of that vast crew that would be known in the pages of American history as the forty-niners.

The *California* was designed to carry 210 passengers, but on that first trip she is reported to have had over 400 jammed aboard. It was a forerunner of what was coming, for from then on men would be streaming to the gold fields by every method they could find. Some would come overland across the plains. Others would arrive by ship, either going around Cape Horn, or crossing from one ocean to another in Central America. And many would never get there; thousands would be buried in hasty graves on the plains and deserts, or would leave their bones in faraway ocean depths.

During 1848 the gold rush had belonged to those living in or near California. News traveled slowly in those days, for although the telegraph had been in operation for four years in the East, it would not cross the continent until 1861. Even when stories of the gold discovery did reach the East and Midwest, they were considered to be tremendous exaggerations. It was only after December 5, 1848, when the President of the United States, James K. Polk, reported the richness of the new gold fields that people became excited. Then, almost overnight, everyone was clamoring to go to California.

Factories manufacturing picks, shovels, boots, and other mining gear found themselves busier than ever be-

On January 16, 1849, the ship Apollo *left New York bound for San Francisco, carrying passengers who wanted to make their fortunes in the gold fields. The* Apollo *did not dock in San Francisco until September 18, 1849. A light line indicates the alternate sea route from New York to San Francisco—with an overland crossing from the Caribbean Sea to the Pacific at Panama—that eliminated the hazardous passage around Cape Horn.*

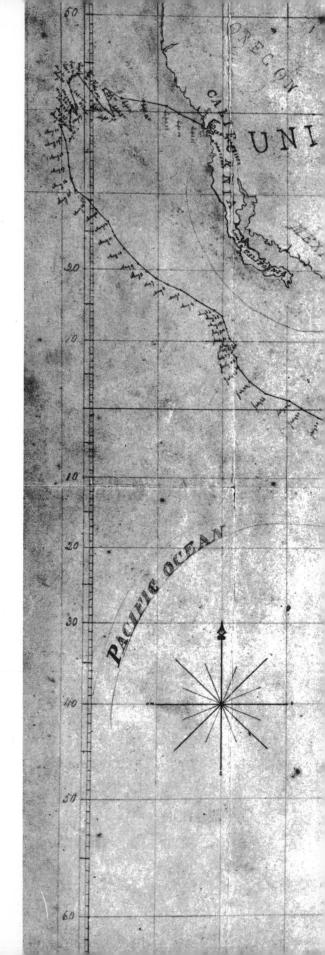

The clipper ship Comet (above), making the run from New York to San Francisco, was battered by a hurricane off Bermuda in October, 1852. The steamship San Francisco (below, center), loaded with 500 passengers bound for California, lost her engines in a heavy sea two days out of New York (December 23, 1853). Through the efforts of the Antarctic (left) and the Three Bells of Glasgow (right) 300 passengers were saved.

fore. In Boston, makers of biscuits and firearms were working twenty-four hours a day. Pawnbrokers did the biggest business in years as men brought in their possessions to get enough money to go west. Farms were abandoned in New England, and in many towns young men set out in groups to go mining. By the middle of January, 1849, one-fifth of the men of voting age in Plymouth, Massachusetts, had left for California.

No story of the gold strike was too fantastic to believe. Men talked of nuggets to be found by pulling bushes up by the roots, and of streams with beds so paved with gold they reflected a yellow light. And everyone was singing a song written by young Jonathan Nichols of Salem, who sailed for California on the ship *Eliza* in the closing days of 1848. It was set to the tune "Oh, Susanna," and one verse went:

> I soon shall be in San Francisco
> And then I'll look around,
> And when I see the gold lumps there
> I'll pick them off the ground.
>
> Oh, Californi—o,
> That's the land for me!
> I'm going to Sacramento
> With my washbowl on my knee.

Ships were suddenly needed in great numbers, and as fast as the vessels came to their home ports, they were put on the California run. Six months after Polk's announcement it was almost impossible to find a ship flying the American flag in most of the harbors of the world. More than seventy ships of New England's great whaling fleet were turned over to the passenger business. Some changes were needed to fit such ships for carrying passengers. Bunks had to be put in spaces meant for cargo, and sometimes an extra hatch or two was cut to provide a suspicion of fresh air during the hot passages through the tropics. But no more changes than necessary were made because men in a hurry to start digging gold would put up with a great deal of inconvenience.

One of the most frightening aspects of the haste to get ships into passenger service was the disregard for human safety. Old hulks with rotten bottoms that had been abandoned as unsafe years before were hauled off the mud banks, patched up, and sent to sea overloaded with eager gold hunters. Many of these floating coffins never survived the trip around Cape Horn.

The traveler by sea had his choice of two routes. He could make the long trip around Cape Horn at the tip of South America. Or he could sail to Central America, cross to the Pacific (usually at the Isthmus of Panama but sometimes at Nicaragua), and take another ship to San Francisco.

Hiram Pierce was one of those who went by the Panama route. Unlike most of the forty-niners, Hiram was not a young man, for he was thirty-eight years old when he left home. He had not been well, and his doctor suggested that a long sea voyage would help him. So Hiram decided to take his sea voyage and get rich at the same time. He closed up his blacksmith shop in Troy, New York, and

Before the last section of the railroad across the Isthmus of Panama from the Caribbean to the Pacific was completed in 1855, the town of Culebra (above) on the Continental Divide was the terminus of the line. Travelers bound for California had to mount donkeys here to complete the last rugged eleven miles of their journey to Panama City.

said good-by to his wife and seven children, and on March 8, 1849, he sailed from New York where he had joined a group heading for the California gold fields.

The ship took them without mishap to Chagres, a muddy, ugly little settlement on the Caribbean side of the Isthmus of Panama, where they boarded a small steamboat that took them seventeen miles up the Chagres River. There they hired five bongos—native flat-bottomed boats—to take them the rest of the way up the river. It was a completely strange world of jungle and river, of monkeys and bright-

colored birds and crocodiles. Hiram Pierce was thoroughly fascinated and put it all down: "The growth of Shrubbery & flower bearing trees & plants surpasses all that can be imagined by a person not familiar with South American senery. . . . The river is shallow & narrow & in many places rappid, up which the nakid negroes pole the bungas or canoes, sometimes wading in the stream, & work verry hard & faithful."

The company reached the jungle village of Gorgona at the end of the second day and made plans to camp about a quarter of a mile upstream. But the boatmen in one bongo rebelled; they had agreed to take their passengers as far as Gorgona, and they would not go even a quarter of a mile farther. Only threats with a pistol made them go the extra distance, and then they refused to unload the baggage until the pistol was flourished again as a means of persuasion.

A train of thirty mules and several native drivers were hired to take the party the rest of the way to the city of Panama on the Pacific. The trail was narrow, and the hot, steaming jungle of brush and vines grew so thick it was impossible to see a dozen feet on either side. The undergrowth cut off all air circulation. Hiram counted forty dead horses and mules stretched out along the route he followed.

The Chagres River, which flowed through the dense jungle at left, was part of the main route across Panama followed by the gold prospectors before the railroad was built.

Captain Edgar Wakeman, shown with his wife in the daguerreotype above, was master of the Adelaide, *one of the many sailing ships that began to carry gold prospectors and settlers from New York to San Francisco after the start of the gold rush in 1849.*

He thought Panama an interesting but tumble-down old Spanish town, but more difficulties awaited him there. "There are perhaps 2,000 Americans now on the Isthmus," he wrote. "The prospect of getting away looks verry dubious.... There are many here that have used up their means, some by gambling & some in other ways, so that they cannot get away."

Some of the stranded travelers were willing to take unreasonable risks to be on their way. Hiram Pierce watched some men outfitting bongos for the ocean voyage to California. "It looks verry hazardous," he wrote. And it was hazardous. One group set out in a lifeboat while he was there, and long afterward he found how the attempt had concluded. After being out two months, the travelers were forced to put in at Acapulco, Mexico, where they left their boat. They finally got aboard a French ship, but it was wrecked by a storm while still at anchor, and twenty lives were lost, including two of the hardy company that had set out so bravely from Panama in the lifeboat.

In Panama weeks went by, and still no ship came. Hiram did some sightseeing, and became very discouraged. His diary sadly records the presence of "2 or 3,000 [forty-niners] on the Isthmus & no Sails arriving to take them off as fast as they arrive. Yet we hope for the best. I think of that Shop in Troy, & that :Wife: whose likeness I look uppon & those dear Children, & my feelings by some might not be called manley."

Meanwhile, crowded, unsanitary conditions began to take their toll. Many Americans were getting sick, and deaths became more frequent. Hiram himself became ill, apparently with malaria; the ailment was to make his life miserable most of the two years he spent in the gold fields.

But finally the company got aboard a ship, and on May 9 they were on their way to California.

"Our mode of living is truly brutish," Hiram wrote. "Our Company is now divided into 2 messes, Starboard and

Larboard. One mess first on one day & the other the next. We form ourselves in two lines when we can, on that small part of the deck that is left clear. And a man passes through with the Coffey. Another with the Sugar. Another with a basket of bread. Another with a botle of vinegar & molases. & then the grabbing commences. We ketch a piece of meat in the fingers & crowd like a lot of Swine. The ship perhaps so careened that you will need to hold on or stagger & pitch like a Drunken man. Many behave so swineish that I prefer to stay a way unless driven to it by hunger."

The ship could make little progress northward; the prevailing winds forced it to the west until by the time it was as far north as California it was more than halfway to the Sandwich Islands. Many ships had to go all the way to Hawaii before they could strike a favorable wind that would serve to carry them to California.

After days of squally weather, Hiram's ship ran into a heavy storm that carried away some of her rigging. For thirty hours she lay at the mercy of the waves, but finally the weather cleared. By then the ship's food was hardly fit to eat, with moldy bread and meat almost putrid, and only one quart of water per person each day.

But the long trip finally ended. On July 26, the ship's seventy-eighth day out of Panama, she entered San Francisco Bay and furled her sails at last. It took the captain two hours to find an anchorage among the 175 ships that Hiram counted in the bay.

Few travelers would have fared better than Hiram Pierce in making the Panama crossing in 1849. Gradually, though, conditions improved. Enterprising men, mostly Americans, set up hotels and saloons along the Isthmus crossing. The mule trains were also better organized, so that the traveler could be fairly sure his baggage would reach the other side without being lost or stolen. By 1851 there was no longer a shortage of shipping in the Pacific, and more and more men took the Isthmus route to the gold fields instead of the long trip around Cape Horn. Finally, in 1855, a railroad was built, and the last difficulties were gone from the land crossing. A year earlier, however, the peak of the gold rush had been reached. Men continued to go to California, but most of them went as settlers rather than in the hope of getting rich in the gold fields.

For the lucky traveler who got a good ship with a good captain and was blessed with favorable weather, the trip around Cape Horn was the most comfortable way of traveling to California. Not many, though, had that perfect combination of circumstances. Most of the gold seekers went on converted craft that were ill-suited for passenger duty. Crossing the equator,

OVERLEAF: *Prospectors hired Indians to pole and row them up the dangerous Chagres River in native boats called bongos. The abandoned bongo at left has been wrecked on a rock. An Indian crewman on the heavily loaded bongo at right has just fallen overboard.*

41

men were unable to bear the breathless, furnacelike spaces below deck and lived and slept on deck. A few weeks later they were in the grip of icy winds at the tip of South America.

The most perilous part of the voyage was the trip around Cape Horn. The winds were frigid and ferocious, and a ship might spend weeks trying to make headway against contrary gales, often being pushed far south toward the Antarctic. A captain who knew what he was doing could save weeks by cutting through the Strait of Magellan, but it was a region of wild crosscurrents and tide rips and unpredictable storms, while on both sides towering cliffs and reefs of wave-swept rock meant certain disaster to any ship that got in trouble.

Even for good, well-handled ships the trip was difficult; and during the gold rush days, when rotten-bottomed ships were being sent to sea under incompetent skippers, the number of wrecks increased alarmingly. Diaries of men who made the trip mentioned over and over again that they had passed floating wreckage, and for many years after the gold rush days, California newspapers kept receiving letters from families in the East, asking for information about the fate of such-and-such a ship which had last been reported in the Atlantic headed toward Cape Horn.

Men who came around Cape Horn were able to carry much more baggage than the Panama travelers. They often brought an odd assortment of articles,

The inset map at right and the 1850 map below show the tortuous route across the Isthmus of Panama before the railroad was built. The journey from Chagres on the Caribbean to Cruces was made by boat or canoe up the Chagres River. At Cruces the trip over the mountains to Panama City on the Pacific was made on donkey back or on foot.

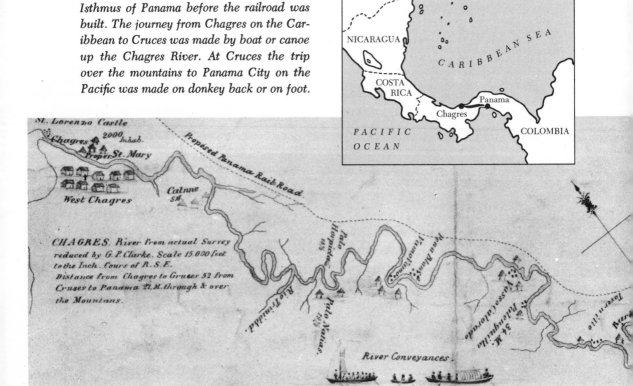

including dredges, ore mills, and strange machines meant to separate gold from dirt and rock by heat or chemical means—all of them completely useless. Some had diving suits in which they hoped to stroll along the bottoms of deep streams and pick up nuggets at their leisure. There were strongboxes to hold gold, and small cannon to repel Indian attacks. Descriptions of San Francisco during the period speak over and over again of the heaps of useless equipment abandoned on the beaches.

While the Cape Horn route soon took second place to the trip via Panama, it remained the only way by which cargo could be brought to California. And because it was important to get goods to California as quickly as possible, the demand for fast ships brought about the period of highest development of the clipper ships.

The clippers were beautiful sailing craft, with clean lines, and long, sharp bows. These speedy vessels cut the average time from New York to San Francisco from a tiresome voyage of six or eight months to a rapid run of 133 days. Clippers such as the *Flying Cloud,* the *Swordfish,* and the *Andrew Jackson* made it in ninety days. But clipper ships were designed for cargo, and while they might carry a few passengers now and then, freight was their main business.

By the end of 1853 the situation in California was less frantic, and the demand for speed lessened. Slower ships could carry more and do the job more cheaply. Soon there was not enough demand for the fast clippers to keep them all busy, and they began to disappear from the seas.

Not all the ships that came to California carried Americans, and not all were American ships. The gold fever reached Europe, Australia, and even China, and men began arriving from all over the world. As a result, by July, 1849, Chinese junks, lateen-rigged ships from the Mediterranean, and the flags of many European nations could be seen among the more than two hundred ships that lay deserted, at anchor, in San Francisco Bay.

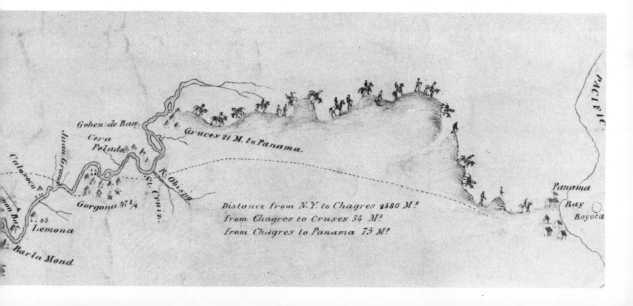

Wagon trains like the one above, shown fording Medicine Bow Creek on the overland trail in Wyoming, had great difficulty in reaching California. Relatively few serious obstacles were encountered on the plains, but the greatly overburdened teams of mules, horses, and oxen had to struggle desperately to pull the wagons through the mountains and deserts.

Overland
to the Golden Rivers

While young men in the East were clamoring to get aboard ships bound for California, those in the Middle West were preparing to go by other routes. For them, the direct way to the gold fields lay straight across the plains toward the setting sun. Soon they were streaming west by the thousands.

There were several possible land routes. Many followed the Santa Fe Trail to Santa Fe, from whence they had a choice of several ways to southern California. Farther north were the Oregon and Mormon trails, which could be followed across the Rocky Mountains before the wagons turned off on several trails either leading to the Columbia River valley or directly to the California gold fields. The trip was full of perils, especially for those who ignored advice. One party on the southern route looked for a short cut

While the big wagon trains were rolling to California, many prospectors—like the foremost man on the belled lead horse—risked making the very difficult overland journey with only a few helpers and pack animals.

and discovered instead an unbearably hot, low-lying desert; many perished, and the place has ever since been known as Death Valley. But even on the well-traveled trails there were hardships, and the routes were soon marked by broken wagons, dead oxen and mules, and lonely graves.

Those who took the overland routes could not leave whenever they wished, as could those who went by sea; their trip was ruled by the calendar. They could not start until the spring was well advanced so that the new grass was green and tall enough to feed the animals. Nor could they wait very long after that time, for they might not get through the western mountains before early snow blocked the high passes. Everyone knew the horrible story of the Donner Party that was trapped in the Sierra Nevadas during the winter of 1845-46, when only cannibalism kept the survivors alive. And so, as the spring of 1849 came nearer, the frontier towns on the Missouri River were filled with crowds of men waiting impatiently to start their journey.

One of the forty-niners who went by land was twenty-year-old Edward McIlhany from the mountains of western Virginia (present-day West Virginia). When he heard a company was forming in the city of Charles Town,

he hurried to join up, but almost did not make it because the roster of seventy-five members had already been filled. At the last minute, however, the membership was increased to eighty, and Edward was one of the five chosen from forty who applied. He paid his share, $300, to the secretary and became a full-fledged member of the "Charleston, Jefferson County, Virginia Mining Company."

The company left on March 3, crossing the Allegheny Mountains by stagecoach, and then going by steamboat down the Ohio and up the Mississippi to St. Louis, where supplies were purchased. From St. Louis they went to St. Joseph, Missouri, one of three towns on the Missouri River which were

starting points for the gold seekers who went overland. The other two were Independence, Missouri, and Council Bluffs, Iowa.

Into these three frontier towns the gold seekers poured all during the late winter and early spring; each steamboat brought a load of eager men who teamed up with others to form companies (if they were not already members of one) and to buy their wagons, mules and oxen, and supplies for the trip. Wagons jammed the steamboat landings and locked wheels in the muddy streets. Most of the men were growing beards, and almost all of them carried bowie knives and pistols. There were many oddities: one man had walked to St. Joseph with his dog

all the way from Maine and was planning to continue on foot to California, and there was a Scotsman who planned to push a wheelbarrow loaded with his possessions all the way to the gold country.

The hotels and boarding houses were filled, and clusters of tents sprang up on the edge of the towns. Gamblers, card sharks, and swindlers appeared. For many men the trip to the land of gold ended right there when they lost all their money and had to return home. For most, though, it was a time well spent because they had a chance to learn something about handling animals and cooking in the open. Many of the men had been clerks and knew little about physical work, and they

badly needed some toughening up for the long, hard trip ahead.

Edward and his partners were also getting ready for this new kind of life. The company bought more than one hundred mules and a number of horses, also sixteen wagons and two large, boat-shaped wagon beds made of iron. These last two were to be used in ferrying rivers and would be cut up in California and the sheet iron used for making rockers — crude hand-operated mining machines used to separate gold from gravel. There was also a small cannon to fight off hostile Indians.

Edward took an immediate interest in the animals, so he was made a mule driver, in charge of a wagon and seven mules. The animals were completely wild and unbroken, and every day they were harnessed and hitched to the wagons and taken out on training trips. Almost every day there were runaways and upset wagons, but luckily no one was hurt and little damage was done.

Then cholera, the most dreaded disease of those days, struck the town. In just one week there were thirty-four

The wagons of the Washington City and California Gold Mining Association—led by J. Goldsborough Bruff—are seen above on July 26, 1849, in the Sweetwater River valley of Wyoming, drawn into a circle for defense against Indians.

Bruff, a skilled draftsman, illustrated the journal he kept of his company's expedition from Washington, D.C., to California with drawings like the one at left, which shows five of the sixty-four prospectors cooking dinner by the trailside.

Two of the Bruff party's wagons are seen in the picture below as they are ferried across the North Platte River of Wyoming on a crude barge constructed of eight canoes lashed together. The horses were forced to swim across the river.

deaths. The terrible fever had spread far and fast. In late 1848, two shiploads of Germans planning to settle in America left Europe. Cholera was aboard, but no one had known it until the craft were well on their way to their respective ports of New York and New Orleans. The disease was carried ashore; by January of 1849 New Orleans was panic-stricken as the number of deaths climbed. The infection got aboard the river boats and spread through the Mississippi Valley, and was then carried onto the plains by the forty-niners. Soon the trails were lined with graves and with polluted water waiting to infect later travelers.

Edward was stricken with cholera but recovered quickly. He gave credit to prompt attention from the company doctor, but his strong, young constitution probably played the biggest part in saving his life. One other man of the company fell ill of the fever and died —the group's first death.

Finally the grass on the plains was high enough to feed the animals, and the long trip began, hurried a little by the fear of cholera. The ferries across the Missouri were soon running from early morning until late at night, but they were sometimes four or five days behind schedule. This same situation was also true at Council Bluffs. Independence was on the far side of the river, and no ferries were needed there. The wagons were carried across the river for weeks, and then, at last, all were gone.

Edward's company was on its way on May 4. The wagon boss had given each wagon a number to indicate its place in line, and Edward McIlhany was Number 7 as he snapped the reins above his team of mules and swung into position. It was a beautiful day, with the new grass tall and green and waving, and the trees along the river in bud. Some of the company looked over the river valley and suggested that it was such a pleasant place that the married men should send home for their families and the entire company should settle down there as farmers. But the rest laughed, immediately vetoed the idea, and started to sing at the top of their voices, "I'm bound for California with a washbowl on my knee."

The plains were an empty place in 1849. Missouri and Iowa were frontier states, and after leaving the Missouri River behind, the emigrants did not see a house again until they came to Fort Laramie, and after that no buildings until they reached the Mormon settlements on Great Salt Lake. The land west of the Missouri belonged to the Indian, the antelope, and the great buffalo herds.

The company was one of the first out of St. Joseph, and for the first few days, as it headed toward the Oregon Trail, the trail was hard to follow. This was a new kind of land for Edward, who came from the mountains. Here, day after day, the wagons were in the middle of a great bowl of sky, moving across the plains toward a horizon that never came nearer. But the weather

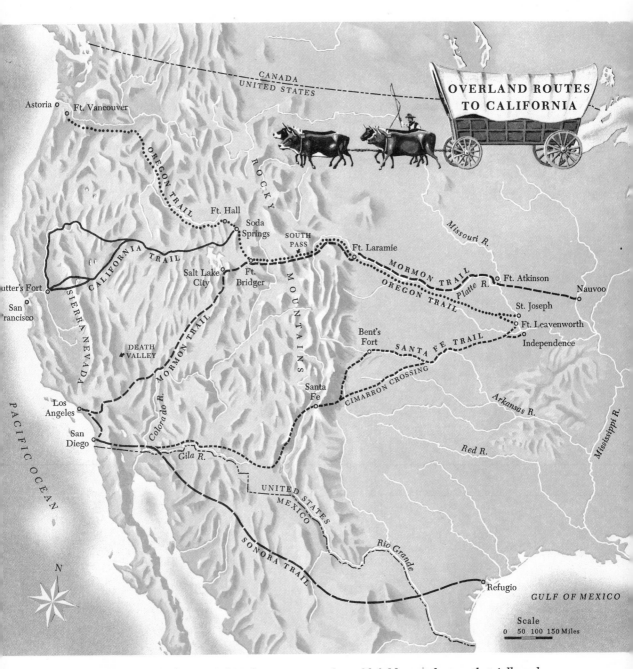

The most frequently traveled of these routes to the gold fields was the one that followed the Oregon Trail from the Missouri to the Rocky Mountains, and from there down the California Trail to Sutter's Fort. In the first six months of 1850 over 39,000 people were recorded as having passed through Fort Laramie, though several thousand more probably passed through unrecorded. The Santa Fe Trail was less frequently traveled by Americans than the northern routes, and the Sonora Trail through Mexico was least traveled of all.

The forty-niner above was painted by John Woodhouse Audubon, son of John James Audubon, the famous painter. Young John went to California via Mexico in 1849.

The steep descent from Mud Lake Basin to High Rock Canyon, Nevada (right), of the Bruff expedition's mule-drawn wagons in September, 1849, was perilous but successful.

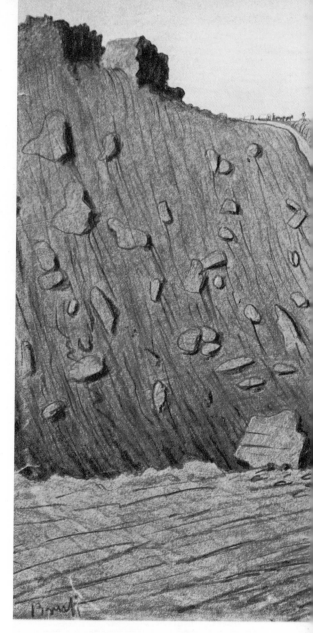

was fine, and everyone was in excellent spirits. Soon they forded the North Platte River, using ten mules to pull each wagon through the stream, and then they were on the Oregon Trail, well marked by years of use.

At another stream a few days later, the water proved too deep to ford, and the wagons had to go two miles upstream to find a crossing and then come back on the opposite bank. Three men decided to save themselves the four-mile trip by swimming across, but it was a bad decision. One of them, a seventeen-year-old boy named Milton, got into trouble in the deep water, and although the others tried to save him, he sank out of sight. His body was re-covered and buried beside the trail with two pieces of board from a wagon to mark his head and feet.

There were other graves on the plains, and there would be many, many more as the wagons kept rolling toward California during the next few years. Most of the dead were victims of disease, and especially of cholera.

Ever since the party had started, there had been Indians along the way,

but when the wagon trains came into the land of the Comanches and Utes they began to be more plentiful. Edward and his partners kept a sharp lookout for trouble, but the red men kept their distance, except for a few who came in to trade deerskins and moccasins. The Indians did not become a serious problem until after the peak gold rush years, 1849-52. Yet in these early years Indians sometimes killed mules or oxen at night, and occasional gold seekers lost their lives, even though the period of bloody raids and ambushes on the plains did not come till later.

As the wagons went west the trail became hilly and often sandy, and the harder pulling was beginning to tell on the mules. They began to grow tired and thin, so their load was lightened by cutting off twelve inches from each

The 1850 photograph at top shows the wagons of a horse-drawn freight train gathered at a stop on the Oregon Trail. These caravans carried manufactured goods and food supplies to the settlers in the Far West. Many towns on the western overland routes grew up at similar freight stations. Because large covered wagons were expensive, settlers and forty-niners unable to afford them often drove small ox-carts like those at center. Before leaving on their westward journey in 1848, the two young Pennsylvanians at left posed for a picture in their miner's outfits.

wagon, except for those made of sheet iron. It was also decided that they were carrying more provisions than were required to get them safely to California. What was not needed—heaps of bacon, flour, and other good supplies—was piled beside the trail and left to the Indians, the coyotes, or other travelers.

Not all parties left their extra baggage for others to use. Many ruined whatever they had to abandon; flour was sometimes doused with kerosene, sugar mixed with sand, clothing set on fire, the barrel of a rifle bent. Earlier pioneers had always tried to help each other, but there were greedy men among the gold seekers who took this way of making it more difficult for others to reach the gold fields.

Now there were many wagons on the trail, and Edward was amused at some of the strange outfits he saw, sometimes a two-wheeled cart drawn by a yoke of oxen, sometimes a smaller cart pulled by a single mule or ox. But he was not at all amused when traffic became so heavy that dozens of wagons were strung out in a long line on the trail. Then they were forced to stay in line, moving all day through choking clouds of dust raised by the wagons ahead. In order to be first in line, Edward's company started rising earlier in the morning.

By the beginning of July, the company had crossed the Rocky Mountains and reached the Green River in Wyoming. Up and down the stream some three thousand other California-bound travelers were camped, resting themselves and their animals before pushing on. Edward's company also rested, making use of the time for cooking, washing, and sewing. On the Fourth of July they selected one of their men as orator and gathered round while he stood on a stump and made a long patriotic speech. Some of the men brought out the small cannon, which had been taken along to repulse Indian attacks, and fired it every time they thought the speaker deserved extra-loud applause. It was the first and only time the cannon was used.

The company also made Sunday a day of rest, unlike many other groups that pushed on as hard and as fast as possible. Edward said that only once did they move on a Sunday, and then only for a part of the day because on Saturday night they had reached a place of scanty grass, too poor to provide grazing for the mules.

The company suffered its third death when a young man named Joe Davis was carelessly pulling a shotgun by the muzzle and the trigger caught in a twig. The blast hit him in the hip, and the doctor could do nothing to save him. He died four hours later. He told his comrades that he did not mind dying so much as the thought of lying all alone out in this great empty land. He was wrapped in a blanket and buried in a deep grave.

When they reached the Humboldt River in Nevada they were beginning one of the most difficult parts of the trip. This was hot, dry desert country.

The few water holes they came upon were so heavy with alkali (mineral salts) that the water was undrinkable. They followed the river, which starts out bravely in the mountains as a large stream, until it gradually sinks away and disappears into the sand. They camped during the day and just before nightfall started across the Humboldt desert. During the night they came to a place called Boiling Hot Springs. Troughs had been set up by earlier travelers into which the hot water could be poured and allowed to cool for the mules, but it was so full of alkali that the animals did not care for it, and several of the men who made coffee with it found it not very good.

They finally came to the Sierras, the last great obstacle before reaching the gold fields, and passed the spot where the Donner Party of emigrants to California (Edward McIlhany insisted on calling it the "Donovan" party even when he was an old man) had been trapped during the terrible winter of 1846 when the living had had to eat the flesh of those who had died. The cabins were still there, although the roofs had fallen in, and stumps cut twenty or more feet above the ground showed how deep the snow had been when the trees were chopped down for firewood. Human bones were still on the ground.

The haul up the mountains was long and hard, and at one point it was necessary to use ten mules to get the wagons up to the crest. At the top the mules were unhitched, a rope was tied around the back axle, and a turn taken around a big pine tree. The wagon was eased down the western slope by paying out the rope. Others had done the same thing before them, and grooves six to eight inches deep had been worn into the tree by the friction of the rope.

The rest of the route was easy and downhill. At a place about forty miles from Sutter's old fort they camped. Here one more member died of what Edward called "galloping consumption." He was the fourth of the original eighty to lose his life on the trip. As for the animals, not a single horse or mule died, or was lost or stolen—truly a remarkable record.

There the company broke up. The mules and wagons were sold and the money divided. Edward McIlhany had grown fond of his mules on the long trip and bought them from the other members of the company.

While the wagons of forty-niners such as Edward McIlhany were still creaking across the plains, one Easterner decided that a quicker and more comfortable way to California was needed, and set out to provide it. He was Rufus Porter, editor of *Scientific American* newspaper. He proposed to build a huge balloon driven by steam engines, which he named the Aerial Locomotive. He had it all worked out on paper; the vehicle would weigh 14,000 pounds; 20,000 feet of spruce rods would form the framework of the huge gas bag, beneath which a passenger cabin of thin boards and painted cloth would be suspended by steel

This 1849 Currier cartoon made fun of the wild rush to get to California. Included among clippers and imaginary airships is a strange dirigible (upper left) that may well have been Rufus Porter's Aerial Locomotive.

wires. According to Porter's calculations, the steam engines would be able to drive the Aerial Locomotive at a top speed of 100 miles an hour.

Porter painted an attractive picture, describing how his passengers would sail in comfort over the mountains, amused by the astonishment of the grizzly bear which looked up as they floated over. But the Aerial Locomotive was never built. Perhaps someone broke the sad news to Mr. Porter that he had overlooked the important matter of air resistance in his calculations and that his balloon could not possibly move faster than a fraction of the 100 miles per hour he predicted.

The conclusion to Edward McIlhany's story is a pleasant one. He tried gold mining for a short time, but found he preferred to haul supplies and equipment for the miners. As time went on he bought more mules, until he was running a train of them between mining camps, and making far more money than most miners did. After several years he returned to Virginia to find that his childhood sweetheart, who had rejected him for another man, was now a widow. He wooed her again and this time won her. Edward lived to be an old man, full of tales about the gold rush and about his long trip across the prairies.

Nuggets and Gold Dust

Gold has fascinated men for thousands of years. It has caused wars, broken uncounted friendships, and been the motive for a multitude of murders. It figures in myths and legends, such as the story of King Midas of the golden touch, and the tale of Jason and the Argonauts and their quest for the Golden Fleece.

Gold also has a number of unique physical properties. Since it will not rust or tarnish, ornaments brought out of ancient tombs gleam as brightly as they did thousands of years ago. It is one of the softest of metals, and is the most ductile and malleable, meaning that it can be drawn out finer and pounded thinner than any other metal. A gram of gold—a piece about the size of a dried pea—can be drawn out into two miles of thin wire. The metal can be hammered to about .00001 millimeter—so thin that it permits a greenish light to shine through.

These things can be done only with pure gold; add just a trace of another metal to it and it becomes harder and more brittle. Other metals are usually mixed with gold to make it hard enough to use.

Gold leaf makes the gilded lettering seen on countless office doors and windows, and it is applied to monuments and the domes of state capitol buildings to keep them gleaming bright and untarnished indefinitely. Microscopically powdered gold is used to produce red stained glass, and, as dentists have long known, the metal is excellent for filling teeth. But only a tiny fraction of the gold in the world is put to such uses. A considerable amount goes into jewelry, but the biggest share of all gold mined is used to pay debts in international transactions.

In its natural state, gold is widely spread over the earth, although usually only in tiny traces. Microscopic bits of it can be found in most rocks, and it is well known that sea water contains gold. But the quantity is so small—about 1/100 of a penny's worth of gold to a ton of sea water—that it will probably always cost much more to remove the gold than it is worth.

Occasionally, a large amount of it is concentrated in one place, as it was in California. In ages past, hot volcanic water apparently brought dissolved gold into cracks in the rocks, until in time veins of quartz laced with threads and bands of the metal were formed. Then, as more time went by, mountains containing these veins were partially worn down by wind and rain and frost until the rocks became sand and gravel, and the bits of gold were set free. Rushing streams carried this debris along, but gold is about eight times as heavy as stones and sand and

is not carried as readily by the current. It is especially likely to drop to the bottom wherever the water slows down behind an outcrop of rock or where the stream widens. Over thousands of years a great deal of gold accumulated in such places in California's streams, and it was there that the gold seekers made their big strikes.

This kind of gold, set free from the rocks and mixed with gravel, is called placer gold. It was a bit of placer gold that caught the eye of James Marshall at Sutter's Mill, and for almost three years the California miners knew no other kind. Then, in the autumn of 1850, when quantities of gold were

discovered still embedded in veins of quartz in the hills, another kind of mining, known as lode mining, began.

The stories that came out of California as to the richness of the gold deposits were usually wild exaggerations. There were no nuggets lying around like hen's eggs waiting to be picked up, and no streams were paved with gold. Most of the men who came to look for wealth would have earned more staying on the farm back home. But strikes were sometimes made that

The daguerreotype above, made in 1850, shows a team of miners operating a small gravel-washing machine called a cradle or rocker (see page 73). The daguerreotype at right shows the operation of a larger mining machine called the long tom (see page 76).

were almost as remarkable as the stories told back East.

One miner on the Middle Yuba River took out thirty pounds of gold in less than a month from a claim only four feet square. Another got $26,000 from his claim on the Stanislaus before it gave out. At the camp of Volcano, miners got as much as $500 in a single panful of gravel. At Durgan's Flat, near Downieville, four men took out $12,900 in eleven days from their claim, sixty feet square; in six months it produced $80,000. Tin Cup Diggings, also near Downieville, was so named because three men who mined there made it a rule always to fill a tin cup with gold each day before quitting.

Just as exciting were the discoveries of large nuggets. Most placer gold occurs as very fine gold dust and flakes, but occasionally a lucky man would find a chunk of solid gold and hold in his hand more money than he had ever seen before. Just up the river from Downieville a nugget weighing twenty-five pounds was found in 1850, and the next year at Sonora a lump of the metal was taken from the edge of a potato patch belonging to a man named Holden; it weighed twenty-eight pounds and was named the Hold-

en Garden Nugget. A tremendous piece of gold, worth $38,000, was uncovered on the Feather River. It weighed 161 pounds—of which twenty pounds were quartz.

Biggest of all was the nugget unearthed at the diggings known as Carson Hill. It was a rich location to begin with. In 1850, a man wandering about picked up a fourteen-pound gold chunk lying right out on the ground, and in the rush that followed, Carson Hill is said to have produced $2,800,-000 in ten months. It was still turning out large amounts of gold in 1854 when the big nugget was found. James Perkins, with four other miners, had come to the end of a discouraging November day and had not taken out enough gold to cover the day's expenses. Four of the five had already quit and gone to the surface; a few minutes later the fifth came rushing up the steep slope of the shaft, almost unable to speak with excitement. The others returned with him, and he pointed out a yellow, gleaming mass he had just uncovered. They brushed the dirt away from the most monstrous nugget they had ever seen. This was the Calaveras Nugget and was the largest single lump of gold ever known

OVERLEAF: *The larger the gravel-washing machines became, the more water their operation required. Systems of water wheels like this one at Cut Eye Foster's Bar on the North Yuba River in California were built to supply water to run the long toms.*

63

to have been found in California. Its weight was 195 troy pounds (gold is measured in troy weight; the nugget weighed about 162 in the avoirdupois pounds used for weighing most other things) and its value was $43,534.

There were also rich discoveries of lode gold. A few Mexicans were crushing gold-bearing quartz by crude means and extracting gold even before the discovery at Sutter's Mill, but they were working on very poor ore, and no one got excited about it. Then, one October day in 1850, a man named George McKnight in the settlement of Grass Valley went out looking for a stray cow. Walking over a hilltop, McKnight stubbed his toe on an outcrop of white quartz and broke off a piece. It was filled with threads of gold. Today the end of that vein has not yet been found. The Empire and North Star mines, now combined under one company, have been working it since 1850 and have taken out $80,000,000. One vein has been followed for 9,800 feet, and the bottom of the shaft is now 1,600 feet below the level of the sea.

Another vein of gold-bearing quartz was discovered just as accidentally. A man with the improbable name of Bennager Raspberry was out hunting one day near Angels Camp, and while reloading his muzzle loader the ramrod became stuck in the barrel. To get it out, he aimed at a squirrel and pulled the trigger. The ramrod missed the squirrel and landed in the roots of a manzanita bush some distance away. When Mr. Raspberry pulled it out, he turned up a small piece of quartz rich in gold. With only his ramrod to dig with, he took out $700. The next day, with better tools, he gathered $2,000 in gold, and $7,000 the third.

Not every story had a happy ending. Consider, for instance, "Gold Lake" Stoddard who had been prospecting with a partner deep in the Sierras. One day he dragged himself, half-starved and with an arrow wound in his heel, into a small mining camp on the upper Feather River. He told the miners how he and his partner had been attacked by Indians. He had somehow managed to escape, but he never saw his unfortunate partner again.

But, said Stoddard, before that happened they had found a lake whose shores were of such richness that lumps of gold were scattered all around it. He promised to lead them there, but nothing could then be done because winter was approaching. By the following spring, the story had spread. When Stoddard finally set out to lead twenty-five men to the wonderful lake, nearly a thousand others followed along as uninvited guests.

Stoddard started out as though he knew where he was going, but he became more and more confused, and at the end of six days he finally admitted that he could not find his way. The most charitable of the men suggested that a landslide had wiped out the lake during the winter, but most were convinced he was nothing but a liar.

Three miners, however, did just as well as though they had found "Gold

Less than two years before this daguerreotype was taken, the busy mining town of Grizzly Flats, California (above), was still an isolated
gold claim in the wilderness. The
sluice and water-wheel system in
the foreground was built to supply
ample water for the gravel-washing
machines known as long toms.
Miners have pitched a tent at right.

Even with the aid of a long tom
(right) a miner still had much backbreaking physical work to do if he
wanted to wash out a large amount
of gold. These men are shown working near Spanish Flat, California, in
1852. Spanish Flat, a boom town
much like Grizzly Flats, was founded
only a few miles from Sutter's Mill.

Lake" Stoddard's nugget-bordered lake. After leaving the search in disgust, they headed back to camp across the hills. In crossing the upper Feather River, they discovered that cracks in the rocky river bed were filled with grains of gold almost by the handful. They were able to gather pure gold without using a shovel; one account says that during the first four days they took out $36,000 without even having to wash any gravel. They tried to keep their discovery secret, but it leaked out, and the rush was on.

This was the location that became known as Rich Bar—and rich it was. Claims were limited to areas ten feet square, which still allowed many miners to become rich. Single pans of dirt containing $1,500 to $2,000 were fairly common, while the highest yield from a single pan is said to have been $2,900. One company of four men took out more than $50,000 in a single day.

Soon, of course, every river and creek with any gold at all had its mining camps. There were hundreds of them, and their names were often rough and usually picturesque. Some, like Iowa Hill, Wisconsin Hill, and Illinoistown, were named after native states. Others, such as Kanaka Flat and Frenchmen's Flat, recalled a nationality group that dug there. Some took their names from incidents that are now a part of folklore: Coffee Gulch, Hog Eye, Loafer Hill, Boomo Flat, You Bet, Cut Throat, and Randy Doodler. Sometimes the names changed. Dry Diggings was so called

The eager miners above had one goal—to find gold nuggets like those at right (displayed with a United States gold piece). The larger nuggets are veined with quartz; the small nuggets, smoothed down by swiftly running water, were found in stream beds.

because of a scarcity of water for washing out gold, but it soon became Hangtown after an incident of miners' justice took place there. Still later, when the gold rush faded and permanent residents, including women, moved in, they thought the name Hangtown was somewhat unrefined, and the town was once again renamed—this time as Placerville, the name it bears today.

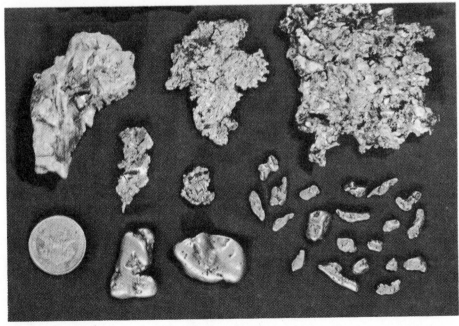

With my Washbowl on my Knee

Digging for gold was hard work—very hard work. Once in a great while a very fortunate miner might find a large nugget or a pocket of almost pure gold dust in a crevice in the rocks, but for most it was backbreaking labor separating small particles of gold from tons of sand and gravel.

When Hiram Pierce (whose journey to California has already been mentioned) had his first sight of gold-mining country, on the American River near Auburn, he was quite impressed by what he found. "Arrived at the diggins," he wrote. "The Senery at the river is wild in the extreme. The water appears to have been brought down or forced its way through a perfect mountain of Granite boulders as they are piled to the height of 100 feet along the banks in some places & are of imence size, 20 or 50 tons."

Much of the gold was found in places like that, along streams and creeks and gullies in the foothills of the Sierra Nevada, where the water,

This prospector of 1850 is using his "wash-bowl," or pan, to take gold out of the sand and gravel of a creek bottom. Even after the invention of far better equipment, the lone prospector still worked streams with his pan.

over uncounted centuries, had cut deep, steep-sided canyons through the rocks. Boulders had tumbled to the bottom, and dead trees were often tangled together to form barriers. Usually, the only way down to the stream was a footpath more suited to goats than men. Many diaries of miners mention the dangerous trails down the sides of cliffs into deep canyons. The newcomer moved fearfully along them, but miners used to them skipped along with complete confidence, carrying supplies on their backs, even at night, in places where a slip could mean a fall of a hundred feet.

The water itself was also a hardship. Running down from melting snow on the mountains, it was icy cold for a great part of the year. Since many miners stood in the water eight or ten hours a day, the lower part of their bodies soaked and cold, it was no wonder that rheumatism was one of the most common ailments in the gold country. This, together with muscles strained from digging, encounters with beds of poison oak, and fingers and toes smashed by rocks, kept gold digging from being the most pleasant of occupations.

Hiram Pierce often told of the

William D. Peck, a gold miner who lived in Rough and Ready, California, sat for this painting in 1852, in his own cabin; the miner's washbowl is lying on the floor at right.

miners' aches and pains, but things were especially bad on the day he made the following entry in his diary: "Though my back is lame, I appear to be the nearest convalesent of anyone here of our party. Daniel Newcom has a verry sore hand caused by poison. Smith has a sort of felon on his hand caused by rubbing on the cradle, & Haskins hands & feet are sore from Scurvey and Sunburnt. My back is lame but I carry dirt & Haskins rocks the cradle. A great many are laid up about us, some with sore hands & feet caused by poison & some by disentary."

But all these discomforts and hardships had to be endured; the miner's real concern was finding gold and separating it from the dirt. (It was always called dirt, even though it was usually sand and gravel.) On his way to California, he had sung many times, to the tune of "Oh, Susanna": "I'm going to Sacramento with my washbowl on my knee." As soon as he reached the gold country he had to learn how that "washbowl" was used, because it was going to be his constant companion until the time he left the diggings.

This implement did, in fact, resemble an old-fashioned washbasin, but its purpose was to separate gold from gravel. When the prospector squatted down on the banks of a stream, he scooped up about half a panful of dirt, and plenty of water. Then he commenced moving the pan so that its contents swished around in a circular motion, meanwhile tipping the pan slightly so that some of the sand and water spilled over the edge as it circled around. More water was added from time to time, and the process continued; at the end the prospector gave a few deft twists and all the sand was gone. Only a few flakes of gold remained in the bottom of the pan.

The principle of the gold pan depends on the fact that gold is about eight times as heavy as a like amount of sand. The metal tends to sink to the bottom while the lighter sand is carried around by the water and spilled over the edge. But there is more to panning than meets the eye. If the miner should swirl the pan too hard, small specks of gold might be lost over the edge. Many miners never did learn the trick; as a result, streams supposedly panned out were gone over again using different methods and often yielded more gold than they had before.

Except at the beginning of the gold rush, the pan was seldom used for actual mining operations. It was highly profitable for the lucky few who struck spots where a single panful of dirt would produce $100 or $1,000, but more often panning would not pay a miner enough for his trouble. The pan remained, though, an important part of his equipment, necessary in prospecting for new claims. An experienced man could wash out a pan of dirt and calculate, from the few specks of gold, pretty closely how much he could take out in a day's work.

But very soon after the discovery at Sutter's Mill, two or three men (who had been miners in Georgia, where gold had been discovered twenty years earlier) began building devices called cradles or rockers; soon, most miners were using them because they could handle many times more dirt in a day with a rocker than they could with a pan. As its name implies, the device was something like an old-fashioned cradle. Its top was covered with an iron plate pierced with holes. The miner would put a shovelful of dirt on the plate, and then rock the cradle back and forth while dipping up water and pouring it over the dirt. The stones rolled off; the sand and gold dust dropped through the holes. On the floor of the cradle two or three strips of wood or "riffles" provided places be-

OVERLEAF: *In the early 1850's, when this picture was painted, small-scale gold mining operations—using devices like the long tom (center) and the gold pan and rocker (lower right)—were still common. Hydraulic mining (upper right), in which hoses were used to wash away large quantities of gold-bearing sand and gravel, was to become more and more common as big companies moved into California and harnessed mountain streams.*

hind which the gold could lodge while sand and water ran out the open end.

Operating a cradle was hard work; it was necessary always to keep dipping water over the dirt with one hand while rocking with the other. One miner compared it with the old trick of trying to pat the top of one's head with one hand while rubbing one's stomach with the other. The rocker brought an important change to the gold country; in the beginning most miners worked by themselves, but when they started to use the cradle they had to team up as partners. It took at least two men to operate a rocker, while three made a very good team: one man to dig, one to carry dirt to the rocker, and the third to work the rocker itself.

By 1851, a device called the long tom had pretty much taken the place of the cradle. It was a wooden trough or flume with a bottom and two sides, anywhere from twelve to twenty-five feet long. At one end it widened out and had a piece of sheet iron with holes for a bottom; under the iron sheet was a flat "ripple box" a couple of inches deep, with riffles across its bottom as in the rocker. The long tom was set at a slight slant and was so arranged that water from the stream ran through it continuously.

The miners shoveled dirt into the long tom, while one of the crew tossed out the larger stones and kept things moving so the dirt was washed through the trough and across the iron plate. The sand and gold dropped through the holes just as they did in the cradle, and the particles of gold lodged behind the riffles while most of the sand was washed away. At the end of the day, the ripple box contained sand and a considerable amount of gold. The final separation was made by panning.

With a cradle, an average of about 100 bucketfuls per man could be washed in a day, while with the long tom 400-500 bucketfuls per man could be handled. And four or five times as much dirt washed meant four or five times as much gold. One or two men could operate a long tom, or as many as six or eight could work on one just by making the trough a little longer. The cradle never did disappear, though, because the long tom needed large amounts of running water, and the cradle could be used where water was scarce.

One other method of taking out placer gold was very common— "coyote hole" mining, in which a shaft was sunk straight down through the earth until it reached the bedrock, sometimes 150 feet below the surface. If there was gold, it was almost always found at the very bottom, for during thousands of years the heavy metal had gradually sifted down through the gravel, dropping a tiny bit each time the earth quivered from the tremors so common in that region, until it came to rest on the rock.

At least two men had to work together, one digging at the bottom of the hole, the other hauling up the dirt

with a bucket and rope. When the rock was reached, the miner enlarged the bottom of the shaft, hoping to cover as much as possible of the claim without causing a cave-in. It was a dangerous occupation. Many miners either did not know how to brace and timber the sides of the shaft properly, or else they skimped on the timbering. Men were all too often crushed and buried when the sides gave way and the dirt cascaded down.

There was very little organized law during the first years of the gold rush, and one would think that there would have been many bloody battles for possession of choice claims. Not so.

The long tom (shown earlier on pages 31, 63, and 67) was a far more efficient device than the gold pan or the cradle. A long-tom team, like the one above, could process hundreds of buckets of gold-bearing gravel in one day.

Instead, the men made their own laws and enforced them—and did so well that the regulations they drew up became the basis for mining law in the United States.

Each camp made its own laws at meetings where all the miners were present and each had a chance to express his opinion. It was perhaps the most perfect example of democracy this country has seen. The laws they passed were very simple—how many

feet of ground a man could claim in the stream and how many on the hillside away from water, how long he could be absent from his claim without losing it, and the like—but they were all that were necessary.

John Borthwick, a Scottish author and artist who wrote excellent accounts of his experiences during the gold rush, described the settlement of one dispute. Six men had a claim in the bed of a creek; in order to get the water out so they could work it they proposed to "turn the river," which meant damming the stream above their claim and then leading the water through a ditch and back into the stream below the claim. But others, whose claims along the ditch would be flooded if the plan were carried out, opposed it.

The decision was left to a jury of miners. A call was sent to all men with claims for two or three miles up and down the creek to meet the next afternoon. Although they did not like to lose time from their own work, about one hundred miners showed up. Each side was allowed to select six of the twelve jurors; then spokesmen for each side made speeches giving their side of the argument, while the crowd sat around on piles of stone, smoking

The ferry at Don Pedro Bar on the Tuolumne River in California was typical of the crude little rafts that were used in the West. It was run from the river bank by a system of pulleys and ropes. The Ferry Hotel across the river was a stop for stagecoaches and was probably a stop for wagon trains as well.

their pipes and listening. After both sides had had their say, the jury went down to the stream to look over the situation, called on some more witnesses for additional information, and then ruled that the company should be permitted to turn the river, but not for another six days—to give those along the ditch a chance to get their gold out before the water flooded them.

"Neither party was particularly well-pleased with the verdict," Borthwick wrote, "a pretty good sign that it was an impartial one; but they had to abide by it, for had there been any resistance on either side, the rest of the miners would have enforced the decision of this august tribunal. From

Using a system of water wheels and flumes, this mining crew of the 1850's (above) is trying to turn the course of a river in California so that they can search for gold in the sand and gravel of the former river bed.

it there was no appeal; a jury of miners was the highest court known, and I must say I never saw a court of justice with so little humbug about it."

During the first years after the discovery of gold, the miner was absolute king. He could stake a claim anywhere, regardless of whether the land was already the site of a farm, a house, or a town. All he had to do was stake out his claim with four pegs, usually decorated with rags or tin cans, and after that the bit of ground was his as long as he worked a minimum amount of time on it, usually one day per month.

There is the story, for instance, of two men in Grass Valley who planned to earn some money cutting and selling hay, then in great demand for horses and oxen. They put a brush fence around a meadow where the grass was good, with the expectation of making at least $400 an acre from the hay. But a prospector climbed their fence, sunk a test shaft, and struck gold. Within twenty-four hours, every bit of the hay ranch was staked out in claims fifty feet square. Not only did the men lose their hay, they did not even get a claim for themselves.

Francis Marryat, an English artist and writer who visited California in the gold rush days, tells a story about the funeral of a miner. At the graveside, the minister was saying a rather long prayer, and some of the kneeling mourners got fidgety and began fingering the dirt thrown up from the grave. The minister became conscious of a buzz of excitement and stopped his prayer to ask what was going on. He picked up some of the gravel himself, looked at it carefully, and then shouted, "Gold! Gold!—and the richest kind of diggings! The congregation is dismissed!" The dead miner was taken to be buried elsewhere, and the entire funeral party, including the minister, staked out claims on the new diggings.

No possibilities of finding gold were overlooked. John Borthwick tells how he once went to the cabin of a friend, only to find deep holes inside where

the floor should have been, and men hauling out dirt. His friend had made an agreement with six miners whereby they would mine the ground under his cabin in return for half of anything they found. So they dug two holes, each about six feet square and seven feet deep, inside the cabin, and took out the gold. Then the holes were filled up again, and life in the cabin went on as before.

With so many thousands of men digging, it was not long before the richest placer deposits were exhausted. There was plenty of gold-bearing gravel left—there still is—but getting the gold out was a job for large companies with plenty of money. Quartz or lode mining also took mining out of the hands of the individual or the partner-

The Helvetia Quartz Mill (below) at Grass Valley, California, was one of the most important mills used to crush ore in the early days of the gold rush. Quartz was brought to the mill in donkey carts like the one at right.

ship of a few men and turned it into big business. It took many men and much machinery to blast the gold-bearing quartz loose underground, crush it to powder, and extract the flakes of gold.

But lode mining also had its human problems and tragedies. Consider the story of Michael Brennan, an alert, happy, and intelligent man who came from New York with his wife and three children to manage a lode mine in Grass Valley belonging to friends. He worked hard and did well, taking out enough gold to pay handsome dividends to the owners.

Then the vein suddenly ended—"faulted," as geologists say. At some time in the past, strains deep in the earth had broken and shifted the solid rock so that there was a break in the vein of gold-bearing quartz. Brennan knew about such phenomena; it was just a matter of seeking about where

The Mokelumne River valley of California, part of which is shown above, was one of the busiest mining areas during the gold rush. When Albertis Browere painted this view of the valley, as seen from Mokelumne Hill in 1852, many small mining settlements had grown up there. Water for the mines was brought down from the foothills of the Sierra Nevada Mountains in the three-sided, inclined wooden aqueduct, or flume, at center.

the vein broke off in order to find it again. His crews searched through the solid rock, blasting tunnels in every direction—now east, now west, now north and south. They could not find the vein. Brennan's partners in the East put up money for a while to keep him going, but they soon became discouraged and refused to give him more. Instead of giving up, as he had every reason to do, Brennan borrowed money on his own name everywhere he could until his credit was gone and no one would let him have any more money or goods.

One Sunday in 1858 his friends became worried because no one was stirring around his house. They entered it and made a shocking discovery: Brennan, in black discouragement, had killed his wife and three children and then taken his own life. Beside his body was a note saying that he was only sorry that he could not also take with him his mother and sister in Ireland because he had been supporting them with gifts of money, and now they would be without anything.

Not long after, others resumed work in the mine and almost immediately found the lost vein only a few feet from where Brennan had stopped. Those who visit Grass Valley today will find the grave of the Brennan family, well-kept and very peaceful.

Downieville, California (left), photographed in 1861, is one of the gold rush boom towns that has survived to the present day. Downieville, on the North Yuba River, outlasted many rich gold strikes and colorful citizens.

A Lady
and
a Miner

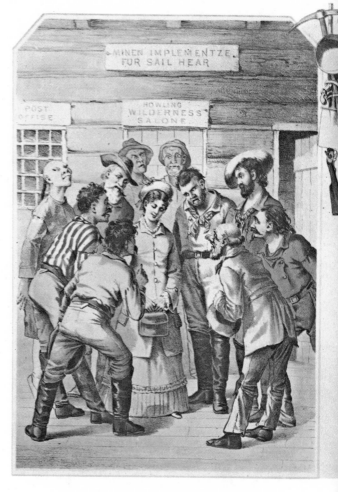

There are many good accounts of life in the California mining camps left by people who were there. The stories of two people who lived through those days are especially interesting. One concerns the forty-niner Hiram Pierce, whose Panama journey has been traced; the other deals with no bearded miner but a young and lively lady. In the name of chivalry, let the lady be the first to speak.

Her name was Louise Amelia Knapp Smith Clappe—Mrs. Fayette Clappe—and, as if all those names were not enough, she had another one. Her childhood nickname had been Shirley,

and in her letters to her sister back in "the States" she referred to herself as "Dame Shirley," the name under which the letters were later published. Although she never had to engage in the grubby, back-wrenching toil of mining, she has left some of the best accounts of life in a mining camp that have come to us.

Dame Shirley came to San Francisco with her doctor-husband in 1849, and for considerably more than a year they lived there. But the damp, foggy winters were too much for Dr. Clappe, who came down with various ailments, and they decided to move farther in-

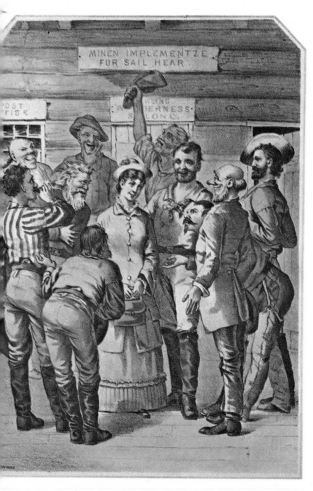

land to the mining country where the air was supposed to be healthier. The doctor selected the camp of Rich Bar, one of the northernmost mining towns on the Feather River, after he was told there was only one doctor there to serve more than one thousand persons. Other doctors must have been attracted by the same news, for while there were only two or three of them when he arrived in the summer of 1851, less than three weeks later there were twenty-nine.

Dame Shirley's friends in San Francisco told her she would detest the wild mining country, and one woman said, as Shirley put it, that "it was absolutely indelicate to think of living in such a large population of men." But, though Shirley described herself in her first letter as "a shivering, frail, home-loving little thistle," she was nothing of the sort, and would not think of missing such an exciting experience. Her husband had gone on ahead to make arrangements for their new home, and when Shirley set out to join him in the fall of 1851, he met her part way. The first thing he did was to get both of them lost, so that they spent more than twenty-four hours and traveled almost thirty miles to get to their next stop only ten miles away. Dame Shirley, after more than twenty hours on muleback and some thirty hours without food, broke down and wept when they were finally within sight of their destination, and begged for nothing more than to be allowed to sleep beside the trail. A good night's rest refreshed her spirits.

Dr. Clappe, who seems to have been somewhat impractical, managed to get them lost again the next afternoon, but this time he had sense enough to camp for the night instead of wandering about in the dark. In the morning they found the trail without trouble, and were congratulated on not being attacked by Indians or grizzly bears.

The final stretch leading to Rich Bar was a five-mile trail down a very steep hill, "along the edge of a frightful precipice, where, should your mule make a misstep, you would be dashed hundreds of feet into the ravine below." The miners they met took the doctor aside and told him it was impossible for a woman to ride down the dangerous trail. They were almost right. Part way down, the girth strap holding the saddle broke, and Shirley tumbled to the ground. By sheer luck, it happened

Mining was a dangerous profession, and some of the wives that followed their husbands to the gold fields were left widows, like the young woman in deep mourning below.

at the one wide, level spot on the whole trail; at any other point she would have gone over the cliff.

Rich Bar, like the other "bars" in the gold country, stood on one of the flat shelves of land which the river had built up, now on one side, now on the other, as it swung back and forth among the hills. "Imagine a tiny valley," Dame Shirley wrote her sister, "about eight hundred yards in length and, perhaps, thirty in width, (it was measured for my especial information), apparently hemmed in by lofty hills, almost perpendicular, draperied to their very summits with beautiful fir trees; the blue-bosomed 'Plumas,' or Feather River I suppose I must call it, undulating along their base, and you have as good an idea as I can give you of the *locale* of 'Barra Rica' as the Spaniards so prettily term it."

The first miners gave Rich Bar its name because of the fabulous amounts of gold they found; some, scooping up gravel from crevices in the rocks, washed out single panfuls that produced $1,000, $2,000, and even more. But it had taken nature untold centuries to concentrate gold in pockets of such richness. They were soon exhausted, and the miners had to work hard for what they panned thereafter.

Rich Bar was one year old when Dame Shirley arrived in September of 1851. There was one street through the middle of the camp, and about forty buildings, "among which figure round tents, square tents, plank hovels, log cabins, etc." They ranged from the lordly Empire Hotel to huts of pine boughs covered with old calico shirts.

Her new home was the Empire, and she gives a fine description of this establishment, which was typical of the better class of hotels in the mining camps. It was the camp's only two-story building, and to add to its grandeur, it had three glass windows—the only ones in the town. It was built of rough planks, and the roof was covered with canvas, as was the front, on which was painted "THE EMPIRE."

No space was wasted on a lobby. On entering the front door, Dame Shirley found herself in a large room whose walls were covered "with that eternal crimson calico, which flushes the whole social life of the 'Golden State' with its everlasting red." One part of the room served as a barroom. In another section was a table covered with green cloth and on it a pack of cards and a backgammon board, ready for an evening of gambling. The rest of the room was a store, "where velveteen and leather, flannel shirts and calico ditto —the latter starched to an appalling state of stiffness—lie cheek by jowl with hams, preserved meats, oysters and other groceries in hopeless confusion."

From this room one went up four steps to a parlor, which contained a round table, six cane-bottom chairs, a rocking chair, and a cookstove, but most impressive, a sofa fourteen feet long and a foot and a half wide, "painfully suggestive of an aching back—of course covered with red calico." There were also red calico curtains.

Emma Johnson (above), the young daughter of a forty-niner who lived in Hangtown, California, posed for this formal daguerreotype in 1850, dressed in her best Sunday bonnet.

Four more steps led to the second story where there were four little bedrooms, only eight by ten feet, with more red calico curtains. What impressed Dame Shirley most were the beds, tremendously heavy, so that she was convinced that they must have been built, piece by piece, on the spot where they stood. The doors were nothing more than slight frames covered with the dark blue cloth called drilling and hung on leather hinges. One of these rooms was to be her home for a number of weeks.

There was also a kitchen and a dining room, which she did not describe.

She summed up her impression of the Empire thus: "It is just such a piece of carpentering as a child two years old, gifted with the strength of a man, would produce, if it wanted to play at making grown-up houses." Yet this flimsy structure had cost eight thousand dollars to build, in a day when that amount of money would have built a splendid house anywhere in the East. But the cost is not surprising when one learns that everything had to be brought in by mule train from Marysville, eighty miles to the south, along a trail that included the cliff-side path on which Dame Shirley had come, at a cost of forty cents a pound for transportation alone.

There were three other women in Rich Bar when Dame Shirley arrived, for this was 1851 and the camps were no longer so wild and rough as they had been in 1849. One of them was the wife of the landlord of the Empire. Another was known as the Indiana Girl—so called because her father owned a nearby hotel, the Indiana House. Shirley, in her room, heard her before they met: "The far-off roll of her mighty voice, booming through two closed doors and a long entry, added greatly to the severe attack of nervous headache, under which I was suffering. . . . This gentle creature wears miners' boots, and has the dainty habit of wiping her dishes on her apron! Last spring she *walked* to this place, and packed fifty pounds of flour on her back down that awful hill—the snow being five feet deep at the time."

The third woman was the complete opposite of the hulking Indiana Girl— a tiny thing named Nancy Ann Bailey, who was said to weigh only sixty-eight pounds. Like the other two, she worked in a hotel, tending bar in the Miners' Home, which she and her husband ran. Shirley was not to know her long because she became ill and died a few weeks after the Clappes arrived, leaving three small children. She was carried to the little hilltop cemetery and laid to rest.

There were other tragedies in the camps. A man was murdered, another died in a shooting accident, still another was killed in a senseless duel. And, of course, the dangerous business of mining took its toll in lives and injuries. Being a doctor's wife, Dame Shirley was no stranger to this side of

Lotta Crabtree (left) was an immediate success when she made her debut as a singer and dancer in California's most isolated gold mining camps at the age of eight, in 1855. The sentimental miners showered the little girl with gold nuggets after her first performance. Lotta was coached by the internationally famous dancer Lola Montez (above), who lived very close to Lotta's family in Grass Valley.

When Lola Montez first arrived in San Francisco in 1853, 5,000 curious men met the beautiful, Irish-born dancer's boat at the docks. After an unsuccessful tour of the mining camps (see page 106) she decided to retire from the stage, and settled in Grass Valley, where her house became a meeting place for visiting Europeans. Lola also bought shares in the Eureka Gold Mining Company.

camp living. When the life of a young miner hung by a thread after Dr. Clappe had amputated his crushed and infected leg, it was she who sat hour after hour for days with the man, brushed away the flies with a pine bough, and forced milk between his lips, a spoonful at a time, until, almost miraculously, he recovered.

The Empire was not the most restful place in the world, with its all-night noises, including the constant profanity that Shirley could never quite harden herself against, and the Clappes moved when they found a small log cabin at Indian Bar about half a mile down the stream.

Indian Bar was a small camp when they arrived, about twenty buildings including those of calico cloth and pine boughs, and the entire level stretch was so covered with the shafts of coyote-hole mines and piles of dirt thrown up from the holes that there was room for nothing else except the buildings and the paths leading to them. The camp was buried even deeper among the hills than was Rich Bar, where the winter sun shone down at least briefly during the day, for Indian Bar was surrounded by peaks so high and steep that the rays of the sun did not reach it at all during several months of the year.

Dame Shirley had poked a great deal of fun at the Empire, but it was a palace compared to the Humboldt, the only hotel in Indian Bar. She described it as "a large rag shanty, roofed, . . . with a rough kind of shingles."

However, there was a good bowling alley attached, the barroom had a floor on which the miners could dance, and a cook who could play the fiddle for their evenings of dancing.

Dancing was one of the ways the men tried to fill the long winter days when snow and rain kept them from working. They rolled ninepins in the bowling alley practically twenty-four hours a day. They played cards, and some of them gambled away all the money they had made the season before. Some spent their time reading dime novels. And on at least one occasion their attempts to relieve the boredom got out of hand.

That occasion was a huge Christmas celebration held in Indian Bar in 1851. The floor of the Humboldt Hotel was washed for the first time, and loads of brandy and champagne were brought in by mule. At nightfall on Christmas Day the miners from Rich Bar arrived, carrying lanterns, and the celebration began at nine o'clock with an oyster and champagne supper, followed by dancing all night. The party kept on getting wilder and wilder. For three days some of the miners did not stop their reveling even to sleep. Shirley's greatest worry was that someone would drown, because drunken men were constantly falling in the river. It was three weeks before the carousing ran completely out of steam, and for some time after, the young miners were very shamefaced when they met Dame Shirley.

Much as she condemned the epi-

The mining families of Sierraville, California, posed for this picture in 1852. The men at far left are shown indulging in two of their favorite pastimes—gambling and drinking.

sode, she was wise enough to understand why it happened. "A more generous, hospitable, intelligent and industrious people never existed," she wrote. But here were a lot of young men, full of energy, confined to a sandy flat "about as large as a poor widow's potato patch." They were unable to go anywhere, had little work to do and little to read, and were deprived of the companionship of young women—until finally they exploded.

She had her first introduction to mining camp justice very soon after her arrival when a waiter at the Empire was accused of stealing from his employer. The miners selected a judge and jury, who held their court in the barroom of the Empire. The evidence against the man was overwhelming, and he was sentenced to get thirty-nine lashes and to leave camp within twenty-four hours.

Shirley was distressed by the cruel punishment, but she was completely horrified a few weeks later by the fate of a man who, with a partner who escaped, was accused of stealing $1,800 in gold dust. The jury deliberated for only a few minutes before sentencing the man to be hanged within one hour. They relented enough, though, to allow him three hours to prepare to die, which he did by drinking himself drunk. Shirley says that many of the miners were sure that the extreme sentence would be set aside, but the man was hanged.

Most of the time, however, Dame Shirley enjoyed life to the utmost. She took walks to Rich Bar, upstream, and to Missouri Bar and Smith's Bar a short distance downriver. She tried her hand at washing out a pan of dirt, while the miners looked on indulgently, and got $3.25, which she sent to her sister. She went boating in the pool above a dam on the river, and when the Fourth of July came, she was called on to produce a flag, which she made from some white cloth, some red calico, and a piece of blue fabric, from the ceiling of the Empire, which had faded to just the right shade of blue. Two new ladies who had come from "the States" with their husbands were at the Fourth of July celebration; the camp was getting downright civilized.

Meanwhile, Indian Bar and the other camps were thriving. Miners joined together into scores of companies to turn the river (or divert its course) into wooden flumes or channels so that they could get at the gold that lay in its bed. When the Clappes had come to the camp the previous autumn, the Humboldt was the only public house there. By the next May it had been joined by the Oriental, the Golden Gate, the Don Juan, and four or five others. "On Sundays," she wrote, "the swearing, drinking, gambling and fighting, which are carried on in some of these houses, are truly horrible." Hundreds of men had come to Indian Bar, not all of them desirable citizens, for here, as all over the gold country, the decent, hard-working miners of the first few years were now being joined by a number of unsavory troublemakers and ruffians.

In the fall of 1852, the bustle along the river came to a complete halt. All the work done to get at the river bed proved to be wasted effort because little gold was found. Shirley quoted the report of one company of thirteen men who, after working for seven months and spending several thousand dollars, ended up taking out just $41.70 worth of gold. The shopkeepers and restaurant and gambling-house owners were penniless because they had extended credit to the miners, and now the latter could not pay.

On November 21, 1852, Dame Shirley looked out through the gray rain on a dismal scene as she wrote her last letter from the camp. Most of the buildings had been abandoned; there was hardly anyone to be seen. "It is said there are not 20 men remaining on Indian Bar, although two months ago you could count them by the hundreds." She and her husband were still there only because their plans to leave

Vaudeville shows, featuring brightly costumed dancing girls and actresses like those at right, reached the gold fields in the late 1850's, when the roads had improved enough to make travel between mining towns possible. Because the female population of the towns was limited to a few hardy pioneers like Dame Shirley, the pretty entertainers were particularly popular with the lonely miners.

two weeks earlier had been frustrated by snow in the mountains, which had kept the mule train out. Finally, when to wait longer brought a growing danger of being snowed in all winter in an empty camp with little food, they went out with the expressman who brought the monthly mail, leaving everything behind but a change of linen.

They got out just in time. A few days later a great storm blocked every trail, and the winter of 1852-53 was one of the most severe those parts have ever known. Dame Shirley never saw the mines again. Back in San Francisco, she parted from her husband a few years later, and taught school there for a number of years. She returned to the East in 1878, and died in New Jersey in 1906 at the advanced old age of eighty-seven.

Although Louise Amelia (Shirley) Knapp Smith Clappe wrote some of the best accounts of mining camp life on record, she was only a bystander. When she washed out a pan of dirt to get the gold in it, it was only for a lark and not because she had to; miners went out of their way to make her life pleasant because to them she was that rarest and most precious of creatures— a woman. For a mere miner, life was not usually so pleasant as hers had been, and more often than not his search for gold ended in disappointment.

One of those who failed was Hiram Pierce. His first experience in the mines was brief. The company with which he

A CHARMING GIRL OF NEW-YORK, IN THE GOLD REGION.

had come to California had planned to stay together, pooling all the gold they found and dividing it equally among the members. Like many other groups that had made similar plans in the East, they soon broke up. It was just as well; when they divided up all the gold they had dug during the two weeks they stuck together, it came to exactly $39.49 for each member.

Hiram had spent those first few weeks on the American River. Now he joined several other men and went to the southern diggings, to fields on the Tuolumne River that were still quite new. They built a cabin against the winter rains, and found that even in sunny California the winters could be very cold and wet. His diary speaks endlessly of high prices, unsuccessful mining, aches, pains, and discomfort.

"Rained last night. Commenced building a chimney. Provisions are verry high. Pork, flour, Coffey, sugar and beans are $1 a lb. Candels $1 a piece. Saleratus [baking soda], $8. The men are, many of them, not getting more than $2 or $3 a day. . . .

"Christmas Day [1849]. I spent most of the day in diggin, though I got but $1. I allso mended my pants & took Supper of our everyday fare, fried pork & hoecake. . . .

The elegant young lady from New York in the cartoon on the opposite page is urging her exhausted gentleman friend to dig harder and find more gold. In the picture below she has found a new and harder-working suitor.

A CHARMING GIRL OF NEW-YORK IN THE GOLD REGION.

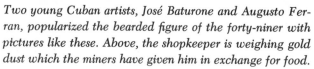

Two young Cuban artists, José Baturone and Augusto Ferran, popularized the bearded figure of the forty-niner with pictures like these. Above, the shopkeeper is weighing gold dust which the miners have given him in exchange for food.

"Jan. 6. Ten months since I left home, & have not made a dollar but am in debt for my board & my health seems insufficient for the task. Oh the loneliness of this desolate region! No Meeting, No Society, nought but drinking & card playing & hunting on the Sabbath. At noon I had a slight chill & in the afternoon a severe head-ache. I felt sore & stiff. . . .

"Jan. 11. I fear the Scurvy. My mouth and gums are sore & teeth loose and legs sore & lame. Still raining. . . .

"Feb. 6. After breckfast worked hard & got $2."

And so his days passed. Things might not have seemed so gloomy to him if only he had been well, but he complained of aches and fevers much of the time, apparently the result of malaria he caught in crossing Panama.

"Mar. 31. I have now spent four months & one half in this place & worked hard & ben diligent & yet lack $125 of paying my board from my earnings, & live most of the time as I would hardly ask a dog to live. I leave this place over $100 in debt."

He moved to Washington Flat on the Merced River, where his luck was hardly any better than it had been be-fore. And there were other things to complain about. Several rattlesnakes were killed, and "grisley bears" were about and had attacked three miners. And the spiders were a frightening size; "I killed 2 tarantlers close to our tent, as big as a small chicken."

Hiram was a religious man, and his idea of a fine evening—and about the only enjoyment he seems to have had —was to have long discussions of religion with another miner. He strongly disapproved of the way many other miners drank and swore and worked on Sunday, and was always hoping they would see the error of their ways.

"Sept. 25. Dug $8 in the forenoon. In the afternoon attended the funeral of Bixby. I was called uppon to act as chaplin. I read the 14 of Job & made some remarks & a prair. It was a Solom Sene, & I hope its impression may be lasting." Hopefully, Hiram added, "I see a marked change in one verry pro-fane man. We bore the ruff coffin along a Mountain path to near our camp where he was decently burried."

But miners went on swearing and working on Sunday, the gold still eluded him, and illness continued to give him trouble. So, as autumn arrived, he did as so many other luckless miners were forced to do and gave up the search. At San Francisco, which he found vastly changed from when he had last seen it, he went aboard the brig *Swift Shure* (as Pierce spelled it), and on October 13, he left California.

He crossed Central America again, but this time at Nicaragua and with his usual luck, came down with fever again. But finally the long trip was over, and on January 8, 1851, almost two years after he left home, he made the final entry in his diary:

"Jan. 8 [1851]. Got into the [railroad] Cars & at 7 P.M. Joined my family with rejoicing. Still weak & feeble with the Chagres fever."

Life in
the Boom Towns

Whenever the prospectors made a rich strike, there were certain other enterprising men not far behind them. In no time at all a merchant would arrive and set up a crude store, ready to sell anything from shovels and shirts to tobacco and liniment. Another early arrival would be a saloonkeeper with his kegs of whiskey and rum. Soon a hotel would be thrown together, a professional gambler would arrive and set up his table, and very shortly, if the gold deposits held out, the camp would become a small city, bustling with business as it served miners who wanted a good many things and had the gold dust to pay for them.

A mining camp was not the place for a poor man. Every miner who ever kept a diary or wrote a letter home commented on the fantastically high prices. One man wrote that he had paid $11 for a jar of pickles and two sweet potatoes, while for one needle and two spools of thread he had to spend $7.50. The price of flour rose as high as $800 a barrel, molasses and vinegar $1 for a 1½-pint bottle, onions up to $2 each, eggs $3 each. And though these prices seem high to us today, they were proportionately even higher in 1849 when a dollar bought much more than it does today.

There was good reason for high

The miners at left, dressed in their best clothes, posed for this photograph in 1854. Men in the mining country did not dress up often because their living quarters were crude and there were few women to impress.

prices. Almost everything had to be shipped all the way around Cape Horn from the East to San Francisco, and then carried to the gold fields by boat, wagon, muleback, and sometimes even by human power. Then, of course, the merchants made sure there was a big profit for themselves. Much more money was made in California by the average merchant than was earned by most of the miners panning for gold. But sometimes the greed of a merchant would go too far even for the miners. An old record from Bidwell's Bar says that "Last week a meeting of miners was called to take into consideration the action of a merchant who had been selling Dr. Stover's California Salve for butter."

The hotels were the busiest places in town. Many of them also contained a barroom and a gambling room. Dame Shirley's impressions of the Empire Hotel in Rich Bar, with its strange combination of crudity and elegance, have already been described. The Empire, however, was considerably better than the average; much more typical was a hotel in Sonora of which an account remains. Although the best in town, it was only a one-story structure, built mainly of saplings covered with canvas. The floors were dirt, and one undivided room served as dining room, parlor, and bedroom. There were tables and benches in the center, while along the walls were five-decker bunks built of crude posts and crosspieces fastened with rawhide.

For bedding there was a small straw

The procession at right was formed in Weaverville, California, in 1860, as part of a German May Day festival. Such towns often seemed as international as cities because of the customs foreign miners brought to them.

mattress two feet wide, a straw-stuffed pillow without any casing, and one blanket. "When we creep into one of these nests," a miner wrote, "it is optional with us whether we unboot and uncoat ourselves, but it would be looked upon as an act of ill-breeding to go to bed with one's hat on."

Although such a building was called a hotel, it was really a crude boardinghouse whose tenants were miners from the nearby diggings. They paid outrageous prices for food and lodging.

John Borthwick has described a meal he had at the United States Hotel while passing through Spanish Bar one day. With sixty or seventy other miners he waited in the saloon until the doors to the adjoining dining room were thrown open; then there was a tremendous rush for places at the benches set on each side of two long tables loaded with ample quantities of roast beef, potatoes, beans, pickles, and salt pork. Not everyone made it, and about twenty unsuccessful men came back into the saloon looking a little sheepish, while from inside there arose a loud clatter of silverware against plates. In an unbelievably short time men began coming out again, many still chewing their last mouthful or picking their teeth with bowie knives.

Then the dining room doors were closed so that the tables could be reset. By the time they were opened again, more miners had arrived, and the same wild rush occurred. Borthwick stationed himself so he could follow in the wake of a broad-shouldered Kentuckian. Once inside, he dashed for a position in front of a large roast beef on a table, and without wasting time trying to sit down, he snatched up a knife and fork and thrust them into the meat. Then, with this for an anchor, he

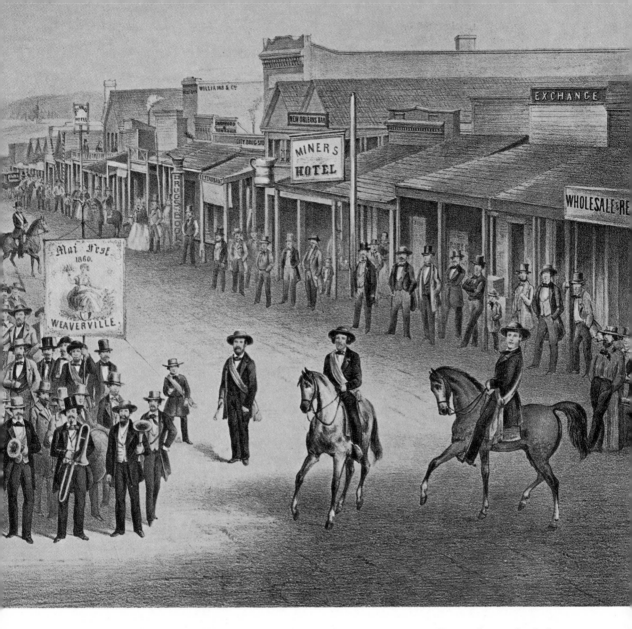

worked his elbows until he had enough room to sit down, while those who had missed getting places this time went back to the saloon to wait for the third table. Borthwick says the diners were all perfectly polite and would pass anything if asked, but they did it with one hand and kept on eating with the other. Their only aim seemed to be to finish as soon as possible and get out of the dining room.

Saloons in the camps did a rushing business, especially at the end of the week, but drunkenness was infrequent. Some men, unable to bear up under the hardships and disappointments of a miner's life, took refuge in drink; but for most the saloon was a place to drop in of an evening and discuss events with friends over a couple of drinks before going to bed.

Strange as it may seem, a saloon owner liked to hire bartenders with broad fingers and thumbs. The price

PORK AND BEANS IN THE GOLD DIGGINS

The cartoon above, showing miners trading gold for provisions, exaggerates the cost of food—which was extremely high to begin with. Merchants in the gold fields often became considerably richer than the miners.

The old and cruel Mexican sport in which bulls were put in an arena to fight a chained grizzly bear (below) had long been performed in California and was viewed by the miners who were eager for new diversions.

for a drink everywhere in the mines was a pinch of gold dust. The miner brought out his leather pouch, and the bartender took as much gold dust as he could hold between thumb and forefinger. The bigger his fingers, the larger the profit.

A story from Shaw's Flat tells of a bartender who found a way to increase his own income. Whenever he took a pinch of gold in payment for a drink, he managed to spill a few grains on the bar. Not much. Not enough for most miners to pay any attention to; but in an evening it could amount to quite a bit. Each time he wiped the bar he swept the bits of gold behind it, and every once in a while he stepped out the back door to get the bottoms of his boots covered with mud from a spring a few feet away. Then, as he worked, he carefully stepped on every bit of space behind the bar so that all the spilled gold was picked up by the sticky mud. After finishing work, he panned out the mud from his boots. It is said that he made about $30 this way every night except on weekends, when his profit was almost $100.

In the early days of the gold rush, a miner's diversions were limited. If he did not want to go to a saloon or to part with his gold by gambling, he had little to do in his spare time. There was an occasional fight between a live bear and a bull, and sometimes there was a horse race where the ground was level enough. As women were a rarity in the diggings, half of the men might tie a bandanna around their arms and be

the "ladies" for an evening of dancing. No one thought it strange to see these bearded "ladies" in heavy boots tripping seriously with their partners through the steps of the polka.

But conditions were changing rapidly in California in the early 1850's, and within three or four years after the forty-niners came, more and more amusements were made available to the miners. As crude trails became roads over which stagecoaches could drive, professional entertainers came

Dinner in a mining-town boardinghouse (below) could often be a dangerous adventure, for the miners—as John Borthwick reported—were fed on a first-come-first-serve basis and had to reach quickly if they wanted to eat.

The miner shown in the center above is per-forming his last important task of the day—weighing the gold dust he has accumulated. The prospectors below are visiting one of their favorite establishments, the saloon.

to the mining towns in the canyons. Before long, every well-established camp with any sort of pride in itself had a theater, singers, variety acts, and theatrical troupes. Even an occasional circus strove for the patronage of the men in the mines.

The miners were enthusiastic when they liked something, but otherwise they were a difficult audience. Lola Montez, who had become famous be-cause of her beauty and her gaudy ca-reer rather than because of her talent, drew tremendous audiences in San Francisco; but when she danced in the gold camps the miners were so bored with her that she had to give up her tour. On the other hand, the miners loved Caroline and William Chapman, a brother-and-sister team who came to

the camps in 1852 and 1853. They acted in plays, or danced, sang, or clowned as the mood struck them, and their delighted audiences threw coins and pouches of gold dust on stage.

The actor Edwin Booth, brother of the man who killed Lincoln, made several trips through the mining camps between 1852 and 1855, usually playing Shakespearean roles to appreciative crowds of miners. But on one trip, fires broke out in Hangtown, Georgetown, Diamond Springs, Nevada City, and Grass Valley while the theatrical group was in each camp, and the miners began to suspect that the players were either jinxed or had a firebug amongst them. At their next scheduled stop, Booth and his fellows were firmly told to keep moving.

Even if there were no women to join in, the Saturday night dance (above) was still an important event in the isolated mining towns. The two shaky miners below had evidently spent their evening out in a local barroom.

The miners do not seem to have been greatly concerned about religion, as the sober and disapproving Hiram Pierce often noted in his diary. There were clergymen among the miners, and they often held open-air services on Sunday, but the men in the diggings were inclined to work on Sunday if their claims were panning out well. Churches were built only after the camps became more settled and numbers of women came to live in them.

Even a clergyman had to be prepared for an occasional rough time from men who did not hold him in much respect. One day in 1851 a minister named Hill came to Shasta in the upper Sacramento Valley and announced that he would hold divine services the next morning. Some of the town's gamblers, probably afraid that an outbreak of spiritual interest would interfere with their business, told Mr. Hill to get out of town. But the next morning he was still around, and showed he meant business by appearing on the balcony of Shasta's largest hotel while a considerable crowd of miners gathered below to hear him.

He had just started to lead his rough-hewn congregation in a hymn when one of the gamblers pushed through the crowd below. "Hey, you," he shouted, "quit that singing and get out

The two soberly dressed boys on burros in this daguerreotype made in the early 1850's were sons of the first gold rush undertaker; his business was in Hangtown, California.

of town." The Reverend Mr. Hill looked mildly at the man, who continued, "All right, get a move on or I'll throw you over the balcony." The minister smiled, and said quietly, "Just help yourself any time."

What happened next happened fast. The gambler ran up the steps to the balcony and rushed at Mr. Hill—and was instantly knocked cold by a hammerlike blow between the eyes. "Now, boys," the minister said, "if any more of you object to hearing the word of God, let him come forward before the service proceeds. I do not like to be interrupted." The miners cheered, tossed their hats into the air, and fired off a few pistols. "Go ahead, Parson, the show's all yours," they called. Mr. Hill preached to a very attentive audience that did not once interrupt him.

One of the heart-warming things about the gold rush days was the willingness of most men to help each other. The story is told of a very down-at-the-mouth young man who appeared at a place where some thirty miners were working. The men looked the stranger over, and asked him why he was so dejected. He had had nothing but bad luck, the young man explained; the claims he had staked all turned out to be worthless, and he was ready to quit and go home.

One miner spoke up: "Boys, I'll work an hour for that chap yonder if you will." And they did. At the end of the hour they turned over about $100 in gold dust to him, and gave him a list of tools, telling him to come back when he had bought them. "We'll stake a good claim for you, and after that you'll paddle your own canoe."

Edward McIlhany, who made the overland crossing to California, told how he came upon a young fellow in Marysville who had lost his legs in an accident. He was unable to do any kind of work, and had no money to get home. Edward and a friend took him to a hotel that had an entertainer, and after explaining the situation to the latter, put the fellow on the stage in a chair. The singer and violinist played and sang a song called "Not Old Dog Tray But Poor Dog Tray" and made a fervent plea for help for the crippled boy. The audience gave generously; more than enough money was collected to send the young man back to the East by ship.

The change in those who came to California was rather startling. These men—so generous to each other—undoubtedly included some of those who, in crossing the plains, had destroyed food and supplies before abandoning them for no other possible reason than a senseless and selfish wish to prevent others from making use of them. John Borthwick noticed the same change in men crossing the Isthmus of Panama. Those coming from the East were rude, argumentative, and complaining, he reported, and they did nothing to help each other or even to do things that would make their own passage easier. On the other hand, those returning from California were thoughtful of their fellow

Sunday was a holiday for the miners; this picture shows how they spent the day. At left a horse race is run, and a miner is celebrating having struck it rich; the men at center are reading aloud; and the man at right is hanging out his weekly wash: his trousers.

travelers, and were willing to put up with any inconvenience necessary for the good of all.

But another side of the picture is not so pleasant. Consideration for other miners did not seem to hold if the others spoke a different language or if their skins were not white. And in a place where there were men from all parts of the world, this attitude led to trouble. In all too many cases, Americans took the arrogant attitude that they had a right to push all others out of the way.

Sometimes differences were settled fairly, even though a bit roughly, as in Rich Bar when, in the summer of 1850, a group of Frenchmen and another group of Americans arrived and started to stake out claims at the same time. A battle threatened for a while, but some sensible person suggested that each side pick its best man and

no doubt that they had every right to stay, but the greed of some Americans was aroused. They spread the lie that the French had raised their own flag and had openly defied the government of the United States. Then, when the mob spirit was high, they forcibly drove out the French and robbed them of their sites.

The same sort of thing happened much more often to Mexicans and to miners from South America, who were called *Chilenos* whether they came from Chile or any other country in Latin America. They not only spoke a different language, they were looked down upon because their skins were swarthy. Americans argued, in an attempt to justify their actions, that since they had just taken California away from Mexico, the Mexicans no longer had any rights there.

Unfair taxes were specially contrived to make things harder for the Latin Americans, and if the claims they staked happened to pay well, they were often driven off them. They were robbed, beaten, and sometimes murdered. It is not surprising that some of them became robbers.

But even the Mexicans and *Chilenos* were held in high regard compared to the Chinese. When the first gold was found at Coloma in 1848, there were only seven Chinese in California; by 1852 there were at least twenty thousand and they were coming in a flood. Many became laborers, and there was always rushing business for a laundryman in the camps. But most

let the two decide who stayed and who went. The fight lasted three hours, but it was a fair one. When the Frenchman lost, his countrymen moved on without further argument. They won in the end, though, because the area they moved to a short distance upstream turned out to be the richest yet found in Rich Bar; to this day it is called French Gulch.

More ugly, though, was the so-called French War near Jackson where a group of French miners had opened up some rich new claims. There was of them answered the call of gold and became miners. Many camps had their own populous Chinatowns where these men from across the Pacific followed their own customs.

They were cheated, swindled, and mistreated. They were permitted to mine only in dirt the whites had already worked over or did not want. Because their ways were so different, they were looked down upon as being not altogether human. When a mule kicked and killed a Chinese at Goodyear's Bar, his countrymen appointed a jury of seventeen men who tried the mule, found it guilty, and ordered it shot at once. It was hard for a young New England Yankee, who had not been beyond the next valley until he came to California, to understand such an incident.

The strangest of all events in which the Chinese were involved were their "wars." The most celebrated of these took place at Weaverville in the summer of 1854. Two tongs or associations of Chinese, sometimes identified as the Cantons and the Hongkongs, had been quarreling for months with many ambushes and beatings back and forth. But when a Canton leader was killed, his comrades immediately issued a challenge to an open battle, to take place in one month.

July was a hectic month. Every blacksmith in Weaverville and the surrounding camps was kept busy making all sorts of strange weapons for the

Chinese: pikes, spearheads with three prongs, and others with curved hooks. As fast as the outlandish spearheads were completed the Chinese fastened them to the ends of 15-foot poles. The law officer in the area, Sheriff Lowe, tried to head off the battle and ordered the blacksmiths to cease making the weapons or else face a $500 fine when the grand jury met. The smiths did some quick figuring, decided they would be well able to pay the fine by the time the grand jury met, and kept on hammering spearheads.

And in the meantime, the Canton and the Hongkong armies drilled and marched and paraded, without once clashing with each other until the day of the battle, when they assembled at Five-Cent Gulch, near Weaverville. Some were armed with the long spears, others carried enormous, two-handed swords, or daggers and shields, and over their heads floated bright banners on long poles.

Two thousand whites had gathered to see the fight, but the armies were in no hurry to come to grips. They marched and countermarched, threatened each other and hurled insults for two hours without clashing. One of the spectators, a Swede known to have a surly disposition when drunk, fired several pistol shots at random into the two armies to stir them to action, and was at once shot in the head and killed by the man standing next to him. It was not that anyone really cared what happened to the Chinese, but the Swede's act was considered unnecessarily cruel and unsporting, and everyone agreed he got what he deserved.

But finally the armies met. Some accounts of the fight say that the Hongkongs had 400 men and the Cantons about 130, but that the latter forced their more numerous enemy into a position where they were split by the spectators, and the portions were defeated in succession. At least seven of the Chinese were killed, and news of the Weaverville Chinese War swept through the mines.

The battle at left did not take place in medieval China, as the weapons and costumes might suggest, but in the mining town of Weaverville, California, on July 15, 1854. By 1852 more than 20,000 Chinese had come to California, and thousands more followed.

Rough
Justice
in the
Camps

Until the miners came, hardly anyone but the Indians had inhabited the gold country. Everything that goes with settlement and civilization was missing, including law and order. There were no police of any kind to protect the miners from robbery or violence, or to arrest men who had committed crimes. What there was of law and justice had to come from the men themselves.

The miners did an excellent job of regulating matters and settling disputes connected with mining. They did almost as well in protecting their lives and property. Very often in boom towns on the American frontier violence and crime ran out of control and no man's life or possessions were safe; but in California the miners took a firm hand and prevented that sort of thing from happening. Theirs was a rough, hard justice, and though there were cases of lynch law, where they hanged men first and asked questions later, the miners usually managed to be fair.

So far as is known, the very first miners' court was held in January of 1849, in the camp then known as Dry Diggings (later called Hangtown, and finally Placerville). The crime was the attempted robbery of a gambler named Lopez. Five men entered his

bedroom at night, and one of them held Lopez at gun point while the others began to gather up his valuables. But the gambler managed to give an alarm even with a pistol at his head, and several other men rushed in and grabbed the five robbers.

The next morning a judge and a jury of twelve miners was quickly selected. The jury decided that all were guilty and ordered that each of them should receive thirty-nine lashes with a rope, and the harsh sentence was quickly carried out. Unfortunately for the culprits, three of them were recognized as men who had escaped after a robbery and attempted murder on the Stanislaus River to the south several months earlier, and they were immediately put on trial again on the new charges.

During the trial the men were still almost unconscious from the flogging, and could not even stand or speak to defend themselves. This time the jury

Joaquín Murietta (left), with his assistant, Three-Fingered Jack, terrorized California during the gold rush with his murders, stagecoach robberies, and cattle thefts. He was finally killed in 1853. The caricature at right ridicules the forty-niner who would much rather steal his gold than dig for it.

did not decide the case; the judge turned directly to the two hundred miners present, and asked what should be done with the three. "Hang them!" the miners shouted.

At least one man did not approve of the proceedings, which seemed to him to be no more than a lynching, and he begged his comrades to give up their plan to execute the three. But the helpless men were taken out at once and hanged. It was probably from this affair that Dry Diggings became known as Hangtown.

This was the·first miners' court of the gold rush, but many others came after and all followed the same pattern: a quick and short trial, and immediate execution of the punishment handed down by the judge. During the early years very few camps had jails, and so there was no way to hold a man until a regular court could meet —even if there had been regular courts. Murder and thefts of large amounts were capital crimes for which a man would be hanged, while flogging was usually decreed for petty thievery. The cruel and grisly punishment of cropping a wrongdoer's ears was also used on occasion.

The miners did not inflict such harsh punishments because they were brutal men. Their only thought was to protect themselves against robbery and violence. Far from enjoying inflicting these penalties, many of the miners were moved to pity. John Borthwick wrote that a man who was guilty of petty thievery was usually such a poor, miserable kind of being that the miners more often felt compassion than anger for him. Sometimes, after a man had had a whipping, a collection would be made and the money given to him to help take him to another camp.

This attitude of the miners is illustrated by the story of a man in Calaveras County who was caught stealing a small amount of gold dust and was promptly given ten lashes. As soon as it was dark, he tried to steal again, and was just as quickly caught. This time he received twenty lashes and was ordered from camp. In the next camp he tried to steal a mule, and once again was taken in the act. He was sentenced to fifty lashes, but when the miners saw the pitiful condition of his back from the two previous floggings, a vote of forgiveness was passed, and the owner of the mule took the culprit to his tent and dressed his wounds.

Although a good many men came to a quick death at the end of a rope, only one woman, so far as is known, was executed by a miners' court. This was partly because there were so few women in the gold country. (In 1850, men made up ninety per cent of California's population.) Miners held women in such high regard that they were not likely to do away with them unless they had very good cause. And prejudice was no doubt partially responsible for the fact that the only woman who was hanged was Mexican.

Her name was Juanita, and she lived in Downieville. There are many ver-

sions of what happened, and the exact truth shall never be known. It is known, however, that on the night of July 4, 1851, a gambler named Cannon was walking home with a friend after a long and vigorous celebration of Independence Day. Passing the small shack where Juanita lived, they stopped just long enough to smash down the door. Some said that Cannon was in love with Juanita and took this odd way of showing it; others claimed that the whole thing was an accident brought on by an excess of alcoholic celebration. Whatever the reason, the two gentlemen apparently came to their senses very quickly after breaking in the door and went home without making any further trouble.

The very next morning—according

It was at Placerville, California (above), in 1849, that the first miners' court issued its first verdict, which led to the hanging of three robbers. The town was established by prospectors who found Coloma too crowded.

to one story—they returned to apologize and offered to pay for having the door repaired. But before Cannon had a chance to say he was sorry, Juanita, a girl of hot temper, plunged a knife into the gambler's heart.

Mining camp justice worked with its usual speed. A jury was selected, the trial conducted, and Juanita was sentenced to death almost before her victim's body was cold. Many of the miners, however, recoiled at the idea of hanging a woman—and besides, the accounts of what had happened were so confused that not everyone was con-

This illustrated set of "commandments" was written in 1853 by J. M. Hutchings, an unsuccessful gold prospector from England who circulated his humorous rules for behavior in the California gold fields among the miners. They found them so amusing that they quickly bought 100,000 copies and tacked them up in their cabins. The pictures represent: (1) two prospectors going into Placerville in 1849; (2) the first baby born on Canyon Creek in 1850; (3) Jack Simpson, a forty-niner, tangling with a ferocious grizzly bear in Eldorado County; (4) the majestic elephant who is supposed to have announced the commandments to the miners (see text at right); (5) miners having a good time in Nevada City, California, in 1849—the miner at center has tried on the hoops for a woman's skirt; (6) a frontier character named Missouri Bill (seated) with two companions on Deer Creek, in 1849; (7) washing gravel on the Yuba River; (8) Major Downie, the Scottish prospector, leading a party of miners and pack mules into Downieville, the camp named after him; (9) prospectors arriving at Sutter's Mill in 1848; (10) a sampling of all the disasters that befell wagon trains, combined in one picture; (11) an Indian massacre on Murderers Bar in 1849, in which twenty prospectors were killed; (12) miners doing their household chores on Sunday; (13) gambling in a Placerville saloon; (14) the general store at the Georgetown mining camp.

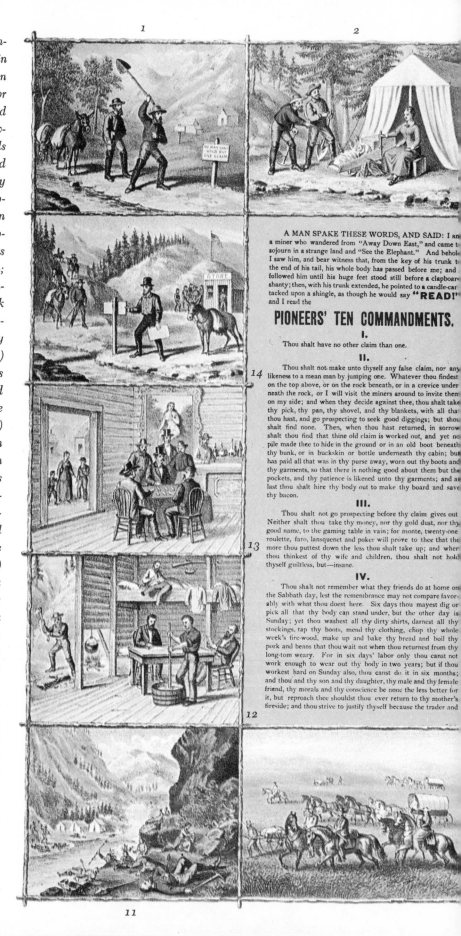

3

4

5

the blacksmith, the carpenter and the merchant, the tailors, Jews and Buccaneers defy God and civilization by keeping not the Sabbath day, nor wish for a day of rest, such as memory of youth and and home made hallowed.

V.

Thou shalt not think more of all thy gold, nor how thou canst make it fastest, than how thou wilt enjoy it after thou hast ridden rough-shod over thy good old parents' precepts and examples, that thou mayest have nothing to reproach and sting thee when thou art left alone in the land where thy father's blessing and thy mothers's love hath sent thee.

VI.

Thou shalt not kill thy body by working in the rain, even though thou shalt make enough to buy physic and attendance with. Neither shalt thou kill thy neighbor's body in a duel, for by keeping cool thou canst save his life and thy conscience. Neither shalt thou destroy thyself by getting "*tight*," nor "*slewed*," nor "*high*," nor "*corned*," nor "*half-seas over*," nor "*three sheets in the wind*," by drinking smoothly down "*brandy slings*," "*gin cock-tails*," "*whisky punches*," "*rum toddies*" nor "*egg nogs*." Neither shalt thou suck "*mint juleps*" nor "*sherry cobblers*" through a straw, nor gurgle from a bottle the raw material, nor take it neat from a decanter, for while thou art swallowing down thy purse and thy coat from off thy back, thou art burning the coat from off thy stomach; and if thou couldst see the houses and lands, and gold dust, and home comforts already lying there—a huge pile—thou shouldst feel a choking in thy throat; and when to that thou add'st thy crooked walking and hiccupping ; of lodging in the gutter, of broiling in the sun, of prospect holes half full of water, and of shafts and ditches from which thou hast emerged like a drowning rat, thou wilt feel disgusted with thyself, and inquire, "*Is thy servant a dog that he doeth these things?*" Verily, I will say, farewell old bottle; I will kiss thy gurgling lips no more ; and thou, slings, cock-tails, punches, smashes, cobblers, nogs, toddies, sangarees and juleps, forever, farewell. Thy remembrance shames me; henceforth I will cut thy acquaintance ; and headaches, tremblings, heart-burnings, blue-devils, and all the unholy catalogue of evils which follow in thy train. My wife's smiles and my children's merry-hearted laugh shall charm and reward me for having the manly firmness and courage to say: "*No! I wish thee an eternal farewell!*"

VII.

Thou shalt not grow discouraged, nor think of going home before thou hast made thy "*pile*," because thou hast not "*struck a lead*" nor found a rich "*crevice*" nor sunk a hole upon a "*pocket*," lest in going home thou leave four dollars a day and go to work ashamed at fifty cents a day, and serve thee right; for thou knowest by staying here thou mightest strike a lead and fifty dollars a day, and keep thy manly self-respect, and then go home with enough to make thyself and others happy.

VIII.

Thou shalt not steal a pick, or a pan, or a shovel, from thy fellow miner, nor take away his tools without his leave; nor borrow those he cannot spare; nor return them broken; nor trouble him to fetch them back again; nor talk with him while his water rent is running on; nor remove his stake to enlarge thy claim; nor undermine his claim in following a lead; nor pan out gold from his riffle-box; nor wash the tailings from the mouth of his sluices. Neither shalt thou pick out specimens from the company's pan to put in thy mouth or in thy purse; nor cheat thy partner of his share; nor steal from thy cabin-mate his gold dust to add to thine, for he will be sure to discover what thou hast done, and will straightway call his fellow miners together, and if the law hinder them not they will hang thee, or give thee fifty lashes, or shave thy head and brand thee like a horse thief with "R" upon thy cheek, to be known and of all men Californians in particular.

IX.

Thou shalt not tell any false tales about "*good diggings in the mountains*" to thy neighbor, that thou mayest benefit a friend who hath mules, and provisions, and tools, and blankets he cannot sell; lest in deceiving thy neighbor when he returns through the snow, with naught but his riffle, he present thee with the contents thereof, and like a dog thou shalt fall down and die.

X.

Thou shalt not commit unsuitable matrimony, nor covet "*single blessedness*," nor forget absent maidens, nor neglect thy first love; but thou shalt consider how faithfully and patiently she waiteth thy return ; yea, and covereth each epistle that thou sendeth with kisses of kindly welcome until she hath thyself. Neither shalt thou covet thy neighbor's wife, nor trifle with the affections of his daughter; yet, if thy heart be free, and thou love and covet each other, thou shalt "*pop the question*" like a man, lest another more manly than thou art should step in before thee, and thou lovest her in vain, and, in the anguish of thy heart's disappointment, thou shalt quote the language of the great, and say, "*sich is life;*" and thy future lot be that of a poor, lonely, despised and comfortless bachelor.

A new commandment give I unto you. If thou hast a wife and little ones, that thou lovest dearer than thy life, that thou keep them continually before you to cheer and urge thee onward until thou canst say, "*I have enough; God bless them; I will return.*" Then as thou journiest towards thy much loved home, with open arms, shall they come forth to welcome thee, and falling on thy neck, weep tears of unutterable joy that thou art come; then in the fullness of thy heart's gratitude thou shalt kneel before thy Heavenly Father together, to thank Him for thy safe return. Amen. So mote it be.

6

7

8

10

9

Both Mark Twain (top) and Bret Harte (bottom) went into the western mining country as young men and found material for some of America's most popular stories. Harte arrived in California in 1854, and while working as a reporter and editor, he wrote his two most famous stories about life in the gold fields—"The Luck of Roaring Camp" (1868) and "The Outcasts of Poker Flat" (1869). Twain went to Viginia City, Nevada, in 1862 to work as a reporter and to try prospecting (with no success). He traveled on to California in 1864 and wrote the most famous of all his western tales, "The Celebrated Jumping Frog of Calaveras County," in 1865.

vinced that Juanita was at fault. Several miners spoke to the jury, asking that it not send the young lady to her death; but the majority supported the verdict, and Juanita was taken out and hanged from the timbers of a bridge across the river. Her last words were addressed to the crowd of miners: "Adios, señores," she called.

As camps became a bit more settled, the miners recognized that they needed some kind of regular court. At first, many of them followed the Mexican system and elected alcaldes, whose duties combined those of mayor and justice of the peace. In 1850 the legislature of the new state of California abolished the office of alcalde and required towns to have justices of the peace instead. But for the mining camps, where things were conducted in informal fashion, it was a change in name only.

Very few of these local magistrates had any experience in the law. Most of them were simply honest men doing their best to be fair and to conduct their courts with something resembling correct legal procedure. But at times some of the homespun judges made strange decisions. The miners of Sonoma elected an alcalde named Sullivan, whose very first case involved a Mexican, Juan Santa Ana, who had been arrested on the complaint of one William Smith who accused him of stealing a pair of leather leggings. Sullivan heard the sentence, found Santa Ana guilty, and fined him three ounces of gold. Then, to make sure he had not overlooked anything, he turned to Smith and fined him one ounce of gold for having made the complaint.

But this was nothing compared to the highhanded way in which Major R. C. Barry made his decisions in Sonora, where he ruled as justice of the peace. Fortunately, he kept a record, and the grammar and spelling were as weird as some of the decisions he describes. One example will show Justice Barry at his best:

"No. 516. This is a suite for Mule Steeling in which Jesus Ramirez is indited fur steeling one black mare Mule, branded O with a 5 in it from sheriff Work. George swares the Mule in question is hisn and I believe so. On hearing the case I found Jesus Ramirez of feloaniously and against the law made and privided and the dignity of the people of Sonora steeling the aforesade mare Mule sentensed him to pay the Cost of Coort 10 dolars, and fine him a 100 dolars more as a terrour to all evil dooers."

So far so good. Major Barry seems to have made a fair decision, and the punishment he ordered appears mild enough. But what happened next did not even faintly resemble either law or equity:

"Jesus Ramirez not having any munney to pay with I rooled that George Work shuld pay the Costs of Coort, as well as the fine, an in defalt of payment that the said one mare Mule be sold by the Constable John Luney or other officer of the Coort to meet the expenses of the Costs of Coort as also the payment of the fine aforesade. R. C. Barry, Justice Peace. John Luney, Constable."

In many of his cases, Barry seemed more interested in collecting the fines and especially his own costs of court than he did in dispensing justice. But in this particular case Sheriff Work had a lawyer who protested Barry's ruling.

The mining town judge seated on the box at right is hearing the case against the murder suspect at far left. The judge must decide if the man should be held for formal jury trial.

Barry concludes the case with this note:

"H. P. Barber, the lawyer for George Work in solently tole me there was no law for me to rool so I told him that I did not give a damn for his book law, that I was the Law myself. He jawed back so I told him to shetup but he would not so I fined him 50 dolars and comited him to gaol fur five days fur contempt of Coort in bringing my roolings and disissions into disreputable-ness and as a warning to unrooly citizens not to contredict this Coort."

But the miners were not helpless in the face of the law and had their own ways of protesting an unpopular decision. At Weaverville the miners along East Weaver Creek, about three miles from town, were doing very well indeed working the rich placer deposits until a company built a dam above their diggings and cut off their water.

This crude miner's cabin (left) is being used as a makeshift courtroom in which to try the case of an accused horse thief. The cabin probably belongs to the man chosen by the miners to act as judge, as the forty-niners ordinarily had no time to build courthouses.

clapped the nine into the camp's small jail and was immediately besieged by the rest of the men involved in the wrecking of the dam, who insisted that they were equally guilty and demanded that they be locked up too. The jail was only eighteen feet square, but all of them—more than one hundred—somehow managed to squeeze in, while Sheriff Lowe went to the county authorities for advice (it was the mid-1850's, and by that time Weaverville not only had such luxuries as county officials but even a small courthouse). The authorities took one look at the bulging jail, and ordered the sheriff to put his prisoners in the courthouse where there was more room.

When suppertime came around, the sheriff realized that he had been completely outsmarted. The local hotels charged one dollar for each meal they served to prisoners, and there was no money in the county treasury. Sheriff Lowe did the only thing he could. In desperation, he threw open the doors of the courthouse and let them go.

"Now get out of here," he ordered, "every mother's son of you, and get your own suppers—or go without."

As they left, the liberated miners paused long enough to give a rousing cheer for Sheriff Lowe.

When their protests got no results they took direct action, marching more than one hundred-strong to the dam and destroying it completely.

The builders of the dam demanded the wreckers be arrested and punished, and the sheriff responded by bringing in nine of the leaders. The hard-working officer of the law was Sheriff Lowe —the same man who tried to prevent the Chinese war in Weaverville. He

Roaring City
on the Golden Gate

When gold was discovered at Sutter's Mill in 1848, San Francisco was a town of nearly one thousand people and about two hundred buildings, many of them little adobe houses built by the Mexicans. For a while, after the gold fever struck the city, it grew much smaller, because nearly everyone left to look for gold, until fewer than one hundred people remained. Then men began arriving from elsewhere, and San Francisco quickly became the wildest, noisiest, most rapidly-growing city in the world, and one of the most wicked.

When ships came through the Golden Gate, not only passengers but often crews and officers headed for the hills seeking gold. When the steamship *California* arrived carrying the first load of forty-niners, her crew promptly deserted, and it was months before another crew was found so that the ship could return to hauling gold seekers from Panama to California, the duty for which she had been brought west. Her two sister ships, the *Oregon* and the *Panama,* arrived a little later, but their skippers had learned from the experience of the *California* and anchored under the guns of American warships in the harbor and refused to allow their men ashore.

The fleet of deserted ships at anchor

The three daguerreotypes above, made by William Shew in the early 1850's, form part of a panorama of San Francisco harbor. Many of these sailing ships have been left rotting at their moorings by crews who have gone off to hunt gold. The eastern owners who had planned to have their vessels make repeated trips from New York or Panama to San Francisco often lost their crews, profits, and ships.

in the bay grew steadily larger. By July of 1849, there were more than 200 craft without captain or crew aboard. A year later there were 526, and many others had sunk or been beached and used as buildings.

It was a strange fleet, a host of ships that creaked as they swung on their anchor chains with the tide, under a forest of masts and spars and a tangle of lines and rigging. Some lay very low in the water, slowly sinking from the leakage through their seams; others were listing at crazy angles, their masts and rigging entangled with those of nearby ships. Many contained rotting cargoes that could not be un-

loaded because of the shortage of labor, and all were completely overrun by an army of huge, fat, vicious rats.

Most of the men who came to San Francisco were interested only in getting to the gold country as fast as possible, but many stayed on, some to become far more wealthy than most miners would. There were merchants of every kind, gamblers, whiskey sellers, draymen, hotelkeepers, thieves, confidence men, shipowners, murderers—almost anything one could name. There were men from nearly every part of the world: New England Yankees and Southerners, Germans, Frenchmen, Englishmen, Peruvians,

The water front of San Francisco (above), lined with hastily constructed frame buildings and beached ships converted into hotels, stores, and warehouses, was painted by the young English artist and writer Francis Marryat, in 1849; in that year San Francisco became a boom town.

Marryat found mud one of the great drawbacks to living in San Francisco (above) in the winter of 1849. There had been no time to pave streets or build adequate sidewalks, and so when northern California's winter rains set in, the whole growing city was engulfed in a sea of thick, black mud.

San Francisco harbor (above) was a forest of ships' masts during the gold rush. Because of the thousands of people arriving there, San Francisco had to be expanded quickly. As a result, flimsy wooden buildings were built. In this type of construction, there was always a danger of fires like the one below, the third to sweep the heart of the city and the crowded water front, on June 14, 1850. The fire did $5,000,000 worth of damage.

Chileans, Mexicans, Kanakas from Hawaii, Australians, and Chinese. And everywhere on the crowded sidewalks the most important man was the bearded, booted miner down from the hills with a pouch of gold—from which everyone in San Francisco was eager to separate him.

It was an expensive city. A twenty-five-cent piece was the smallest coin in use. Shipmasters were offering $30 a day to men to unload cargo, but there were few who were willing to work for those wages. Whiskey sold as high as $40 a quart, and the man who preferred something not so strong still had to pay one dollar for a bucket of water fit to drink. Rents were fantastic; gambling houses paid as much as $100,000 a year for a roughly constructed building. Those who needed a bed for the night paid eight dollars

San Francisco (above) had become a sprawling, busy city in the early 1860's, when this daguerreotype was made, with little resemblance to the village of 800 people which it had been in 1847, just before the gold rush.

or more for the privilege of a narrow cot jammed against others in one large room without partitions.

Solemn, sober Hiram Pierce did not approve of San Francisco when he first arrived in July of 1849. "Went ashore & found such a state of things as all-most to intoxicate a person without giving 50 cts. a glass. Money seemed of no account. . . . San Francisco is a miserable dusty dirty town of some 5,000, out of every kindred tongue and people under heaven."

Hiram was back again very soon after his first attempt at mining failed, and he was astonished at the change. "San Francisco, I judge, has fulley doubled since I left here 6 weeks hence." He was wrong, of course, because the city had not quite doubled, but even so, it is remarkable that its change in such a very short time could so impress him. Almost every miner who returned to the city after an absence remarked on the same thing; growth and change had been so great that they recognized almost nothing from their last trip. By 1850 San Francisco contained nearly 25,000 people; and the population rose to 50,000 within the decade.

It was a very flimsy city. Its buildings were hastily put together, of rough lumber and canvas like those in so many of the mining camps. Tents were common, and there was no shortage of tenants willing to pay fantastic rents for any kind of cheap makeshift shelter. The town grew not only over the hills on the landward side, but also out over the bay. The bottom of the bay sloped out gradually, and enterprising men, who wanted good business locations on the water front, put up buildings on pilings with wooden wharves to serve as streets. In 1851, John Borthwick observed that the streets and buildings had been built for almost half a mile out beyond the high tide mark. (On that same visit, Borthwick spoke of a small cottage of sun-dried adobe bricks on the town plaza, the only house still remaining from the days before the gold rush.)

Some ingenious men saw the fleet of deserted ships as the answer to their need for buildings and bought them for very little. They were hauled as close to shore as possible at high tide, and beached. Shorn of masts and rigging, with doors and windows cut in their sides, they were much sounder buildings than most in the ramshackle city. As San Francisco spread over the bay on wooden stilts, other buildings grew up around these converted ships, and the plank streets presented a strange sight, with rows of ordinary buildings interspersed here and there by a ship hulk, bow to the street. At one time, 148 beached ships were in use, serving as stores and warehouses. One was a bank, another a Methodist church, and the *Euphemia* became the city's first prison.

San Francisco was a filthy city. It grew so fast that it would have been impossible for even a hard-working city government to keep up with the need for sewers and pure water—and the government of San Francisco was neither honest nor hard-working. The streets became seas of mud during rainy spells, in a manner almost impossible to imagine. Dogs and even drunken men drowned in the mire, and more than once horses sank so deep that they could not be pulled out and had to be shot.

All manner of things—including un-

OVERLEAF: *On July 11, 1851, "English Jim" Stuart, a member of the notorious Sydney Ducks, was hanged by the Vigilantes aboard a ship in San Francisco harbor. He was accused of murder and robbery. Another member of the Australian gang, John Jenkins, had been hanged by the Vigilantes a month earlier.*

wanted ship cargoes—were thrown into the almost liquid streets to provide some sort of footing. A shipload of cookstoves disappeared under the mud, and so did quantities of tobacco, flour, and similar foodstuffs. It is no wonder that the mud became black, rotten, and stinking.

It was almost unbelievable that San Francisco was not wiped out by epidemics, but the city was usually free from pestilence. The one exception was an outbreak of cholera in October of 1850, brought from Panama. Five per cent of the city's residents (one out of every twenty) died, but it was much worse inland in Sacramento where fifteen per cent were taken. The epidemic died away with the coming of cool weather; cholera struck again, but less severely, in 1852.

Conditions improved somewhat as time passed. In 1852, John Borthwick noted that all except the least-used

The brig Euphemia *(below) became San Francisco's first jail in August, 1849. It cost the city $260 to clean up the ship and fit it out with balls and chains and handcuffs.*

streets in the central part of town had been paved with planks. Farther out, though, things were still bad, and it was common there to see six horses, in mud up to their bellies, struggling to pull a half-submerged cart.

The wood and canvas architecture of San Francisco made the city a tinderbox, and six times in a year and a half it was struck by devastating fires. The first one came on the morning of the day before Christmas in 1849. There was no fire department and, though men hastily formed bucket brigades, there was no stopping the fire. Explosions of stores of blasting powder meant for the gold mines added to the terror, while lawless men roamed through the heat and confusion, looting whatever they wished.

Rebuilding began immediately, and soon the city was bigger than ever; then a second fire broke out on May 4, 1850, doing much more damage than the first one. And again men put up more wood-and-canvas buildings and returned to the frantic business of making money until a third fire, on June 14, spread to the part of the city over the bay, where the flames roared under the plank streets in tremendous drafts caused by the heat, to burst forth suddenly and devour people almost before they knew the fire was near them.

Once more the city was rebuilt. A fourth fire struck on September 17, 1850, and eight months later on May 4, 1851, a fifth almost completely wiped out the city. It spread with terrible speed, roaring through the dry build-

ings so fast that many people were engulfed as they ran. The fire companies did what they could but were almost helpless. One of the buildings saved was a warehouse whose owners hired men to pour the eighty thousand gallons of vinegar it contained over its roof and sides. And six weeks later, on June 22, the sixth fire burned many newly constructed buildings as well as those spared by the previous fires. It was the last of the great fires, but it was more than enough: at least 25 million dollars in damage had been done besides the huge toll taken in human lives.

There had been looting during each of the fires, and it was believed, with good reason, that the flames had been deliberately set for just this purpose.

The 1856 parade of a San Francisco volunteer regiment (above) was probably staged by the second Vigilante committee in an effort to frighten the criminal element.

San Francisco was almost completely lawless. Unscrupulous men from all over the world had come there. They stayed in the city where they could rob, assault, set fires, commit murder, and resort to any other crime with no fear of being punished.

One of the very worst groups in the city was made up of men from Australia. For many years Great Britain had used that land as a prison colony, but when the gold rush came, that nation sent many Australian convicts to San Francisco in order to be rid of them. Most of them settled in a part of the city that became known as Sydney Town, because many of its unsavory citizens came from Sydney, Australia. Criminals of other nationalities also drifted there, and it became a disreputable, brawling district where an honest man did not dare to go. Its residents were nicknamed Sydney Ducks, and whenever they grew troublesome honest people said: "The Sydney Ducks are cackling."

Even before the Sydney Ducks had become a problem, San Francisco's decent citizens had made an attempt to bring about law and order. Their target was a group of men who called themselves the Hounds and were the remnants, largely, of a New York regiment called Stevenson's Volunteers, which had come to California to fight the Mexicans and remained after the war. They roamed the city doing pretty much as they pleased. A Hound would walk into a saloon and demand whiskey, or go into a store and pick out a hat or shirt that pleased him, or enter a restaurant and eat what he wanted—giving nothing in return but abuse. Frightened by the Hounds' threats that they would wreck their places, the merchants seldom protested.

On July 15, 1849, the Hounds took it upon themselves to drive all *Chilenos* out of the city. They started this project with a parade with banners, fifes and drums, and officers in uniform; then they went on a tour of the saloons, demanding liquor and smashing bottles and mirrors if the owner seemed at all slow about giving them free drinks. Finally, with their courage fortified, they raided the *Chileno* settlement, tearing down or setting fire to the tent city there, and beating anyone they caught.

The other citizens of San Francisco were not much concerned about the *Chilenos*, whose settlement was a crime-ridden slum, but it did occur to them that if the Hounds found they could wreck and loot one part of the city with no one to stop them, they might later do the same thing in other sections. It was time, men said, to stop the Hounds. A leader of the citizens was shrewd Sam Brannan, the same man who had brought a bottle of gold and showed it in San Francisco streets in order to inflame the city with gold fever. Now he was one of the foremost financiers in the city on the bay.

The citizens immediately started rounding up the Hounds and put them on trial only three days after the raid on the *Chilenos*. Their leader was sentenced to ten years in prison, and others were given lesser terms, but there was as yet no prison to keep them in, and in the end they went unpunished. They had been thoroughly frightened, however, and were never again a problem. Some fled to the mining camps where they found justice much tougher than it was in the city; a few settled in Sydney Town.

But there were other hoodlums and cutthroats, and lawlessness increased. There was an average of at least one murder every day, and countless cases of robbery, assault, and arson. It was

The United States Post Office in San Francisco (above), seen here during the gold rush, received all the mail for prospectors working in the northern California mining area. Huge lines formed when monthly deliveries arrived.

Vigilante headquarters in 1856 (below) was nicknamed Fort Gunnybags because of its heavy sandbag, or "gunnybag," fortifications. This picture is stamped with the Vigilante seal, which bears an accusing eye at its center.

the great fire of May 4, 1851, that stirred the people to action. The Sydney Ducks had boasted beforehand that there would be a fire on that date, the anniversary of another conflagration a year before, and when the fire came they were on hand, looting, hauling out valuables in the heat and confusion, utterly arrogant in their sureness that no one would dare to stop them. Two of the town's principal citizens turned to Sam Brannan for help in curbing the Sydney Ducks. The three men drew up a list of trustworthy citizens and summoned them all to a meeting at the California Fire Engine House the next day.

At the meeting, Brannan urged his audience to take action against the criminals, since dishonest policemen and city politicians would do nothing. The citizens present, 180 strong, organized a Committee of Vigilance. Their bylaws were short and direct. First, the Vigilantes would arrest, try, and punish men for murder, robbery, and arson. Second, every person known to be a criminal was to "leave this port within five days of this date" (June 9, 1851). And third, a committee was set up to look over all immigrants coming to San Francisco and to reject those who did not prove likely to become good citizens.

The signal for the Vigilantes to gather was to be two taps on the firehouse bell, and it was sounded almost immediately. A Sydney Duck named John Jenkins had stepped into a shipping office while the clerk was absent, and in broad daylight, walked out with the money strongbox on his shoulder. With complete calm and insolence he pushed through the bystanders, climbed into a boat, and rowed away. Volunteers in boats took after him. As they caught up he threw the strongbox overboard and laughed at his pursuers, daring them to prove he had stolen anything. When he was brought before the Committee of Vigilance in the firehouse he was still boasting that he would go free.

Sam Brannan was judge during a trial that lasted three hours, in order that no one could claim that Jenkins did not have every chance to defend himself. But the thief still refused to take things seriously, even when he was adjudged guilty and sentenced to be hanged. He was certain that he would be rescued by his fellow thugs and cutthroats. The committee was worried, too. But if there were ruffians planning to make trouble, they were overawed by the discipline and businesslike manner of the Vigilantes, for they did nothing.

When he was led out of the courtroom, Jenkins was still declaring that he would never hang. But when a noose was placed around his neck and still his friends had made no move to rescue him, he began to suspect that the impossible was about to happen. A large number of men seized the other end of the line in order for the responsibility to be widely shared, and then, at a given signal, John Jenkins was jerked into the air.

The city officials, dishonest and often in league with criminals, did not like what the Vigilantes were doing. A coroner's jury charged Sam Brannan and eight other members of the committee with responsibility for the execution, but at once the rest of the 180 signed a statement that they were equally responsible. The officials could do nothing against so many.

Before their work was finished, the Vigilantes had publicly hanged three more men and ordered other wrongdoers to leave town. The power of the Sydney Ducks had been broken.

The Committee of Vigilance remained active for two years, although it was seldom needed again. It disbanded in 1853, deciding its work was done, but the dishonest politicians and criminals became active again, and in 1856 the Vigilantes once more organized. It would not be true to say that they made San Francisco a law-abiding place, because for many years, long after the gold rush was over, it was a city of very poor reputation. But they did bring it a degree of law and order when it was most needed.

Isaiah Lees, seated at far left in the carriage below, holding a bottle and glass, devoted a forty-four-year career in the San Francisco police force—begun in 1852 during the gold rush—to suppressing the city's criminals.

Oh, Californi-o!

There had to come a time when all of California's rich placer gold deposits would be worked out. Although there is a tremendous amount of territory in the California gold country, it took only a few years for the tens of thousands of men swarming through the hills to find the valuable deposits and wash the richness out of them.

As early as the end of 1850, ghost camps began to appear in areas where all the gold had been removed from the nearby hills. One of these was the deserted village of Washington on the Yuba River. Even at that early date its brief life was over, leaving empty cabins, a large hotel that would never again provide bed or meals, and several melancholy, vacant stores. A few

miles farther up the river was the abandoned camp of Canyonville.

Just as hard-rock mining, with its need for many complex machines and much labor, was far too expensive for the independent miner, so did placer mining become too expensive for him once the rich deposits were gone. For in order to make placer mining pay in the areas of poorer dirt, a great deal of earth had to be washed—a process that required large quantities of water.

The portrait above, called A Fortune Made, *shows a content, older miner who has already struck it rich. The two daring young miners at right, their careers ahead of them and ready for any adventure, are seen riding down a mountain flume in a homemade boat.*

Great networks of costly flumes and canals had to be built to take the water where it was needed, and water companies were formed which brought the water and sold it to hundreds of mines. Some of them went into the High Sierras, a mile above sea level, to construct reservoirs.

Flumes on high stilts carried water across deep canyons. Others were fastened somehow to the faces of sheer cliffs hundreds of feet high. Ditches were sometimes chopped through the solid rock. In one place in the northern mines, three water companies joined together to dig a way through a great ridge of rock. Men dug and blasted, and dug and blasted again, exploding ton after ton of powder. In one area a canal eighty feet deep was burrowed through a mile of solid granite.

Many men joined together to work these great systems, all dumping their dirt into long sluiceways and later dividing up the gold they took out of the baffles. In Dutch Flat the sluices were more than a mile long.

In the search for ways to get more and more dirt through the sluices, two young men came up with an invention in 1852 that was to make it possible to work very lean placer deposits—and which would, for a while, bring a completely new kind of mining to the gold fields. Joe Wood, a minister's son from Ohio, and John Payne, a carpenter from Maine, thought that water could be made to do most of the work. They obtained sufficient pressure by bringing water from a dam at a consider-

able height above them and down through a square, boxlike pipe of planks—the only kind of pipe available to them. To the end of the wooden pipe they fitted a crude canvas hose. Then they opened the gate at the dam, and the water rushed down through the box pipe and out the nozzle. Joe Wood played the stream back and forth across the face of a gravel bank. It collapsed under the force of the water, and the gravel was washed down into a sluice below. It was a crude device, but it allowed them to wash far more dirt into the sluices than they ever could have shoveled by hand.

The early makeshift equipment was soon improved. Iron pipe took the place of the wooden box, and the nozzles of the hoses were developed into great machines, called monitors, that stood on a firm base on the ground. All the operator had to do was work a handle to direct the stream back and forth, up and down, while the pile-driver force of the water, flowing as fast as 100 miles an hour, tore apart the hills of pay dirt. Boulders weighing hundreds of pounds were tossed through the air like pebbles, and rivers of mud flowed through the sluices, leaving their gold behind.

Even today, the results of these workings are awesome. The lands look like the Bad Lands of South Dakota or the eroded country of the Southwest; great spires, cliffs, peaks, and towers of earth carved by the water soar high in the air. In North Bloomfield there remains a pit so monstrously

The hydraulic miners above, working in Nevada County, California, in 1866, are "piping the bank"—the hosing operation by which entire mountain sides were often cut away.

Hydraulic mining, as seen below, was so cheap and simple a process that it allowed the washing of gravel that contained as little as five cents' worth of gold in a cubic yard.

large that it is hard to believe that it was made by nothing but streams of water; it is one and one-eighth miles long and 350 feet deep at one end, and 550 feet deep at the other. Out of this hole came $4,000,000 in gold.

But hydraulic mining also had its disastrous results. The millions of gallons of water pounding out of the monitors rolled down into the rivers, carrying their huge loads of earth. This thick mixture of mud was called "slickens"; like a smothering tide, it filled the rivers and spread over farmlands. The Yuba River, once crystal clear and almost thirty-five feet deep at Marysville, became so choked that it was almost level with the streets, and even the protective levees could not prevent the town from being flooded during high water. Thousands of acres of rich farmland were buried, and one orchard was covered with slickens twenty feet deep. Debris even started to settle in the upper part of San Francisco Bay.

For about thirty years the miners ignored protests, acting as if things were just as they had been in the early days when mining laws were superior to all others in California. But great changes had taken place, for early in the 1860's agriculture had become more important than mining. The farmers fought back, passing an anti-debris law that required miners to pay for any damage caused by the dumping of waste on farmlands. The mining companies refused to give up what they considered their right to pour

slickens over the countryside, and the anti-debris law was taken from court to court, until in 1884 the state supreme court of California ruled that the law was constitutional.

That put an end to hydraulic mining, except in the area around Weaverville on the Trinity River in the northernmost fields, where there was little farmland for slickens to ruin.

Of course, by the time hydraulic mining was outlawed, the gold rush had been over for many years. Even as early as 1851 a large part of the rich placer diggings had been pretty well worked over, though there was still

Since hydraulic mining required the washing of huge quantities of gravel, systems of long toms (above) were built to lengths of more than 300 feet. The water supply developed for the hosing operations provided ample water for the long-tom system as well.

In the 1850's the Laird and Chambers Diggings (below) in Nevada County, California, was the site of one of the biggest hydraulic mining systems of the day. Tremendous pressure was built up in the hoses as the water rushed down from the high flumes overhead.

plenty of gold for an independent miner who was willing to work for it.

Then, in the spring of 1859, something happened which can be taken as bringing down the curtain on the great days of the California gold rush. At that time, two men started working a small claim across the Sierra Nevadas in what was soon to become the state of Nevada. It yielded a considerable amount of gold, but the metal was mixed with some other substance which they called "blue stuff," from which it was very difficult to separate

a deposit of almost unbelievable richness! The news, of course, traveled rapidly. A majority of prospectors gave up scratching at the thinning placer deposits of California and headed for the new bonanza—the fabulous Comstock Lode.

Today the deep-rock mines still operate in California. So do modern dredges that float on their own small lakes, digging and extending a pond in front of them and filling it in behind. As they work their way through endless beds of gravel they extract the tiny bit of placer gold each scoopful contains. These methods produce enough metal each year to keep California in fourth place among all the fifty states as a gold producer, next after South Dakota, Utah, and Alaska. But along the rivers once so full of placer gold there are only memories.

The old records list 546 camps and diggings of former importance during the gold rush. Today 295 of them have completely disappeared; even their locations are forgotten. A few of the remaining 251 have become busy cities, a few more cling to life as sleepy little hamlets, but with most, all sign of past human activity has gone, and only the name remains.

Hornitos is a fairly typical example of what happened to the gold camps. Once it was a place of fifteen thousand

the gold. A curious rancher took a sample of the ore across the mountains to California and had it assayed, or tested. The ore contained $876 worth of gold to the ton, which in itself made the claim very valuable, but there was also $3,000 worth of silver to the ton—

This vast gold and silver mine outproduced all of the other mines in the Comstock Lode.

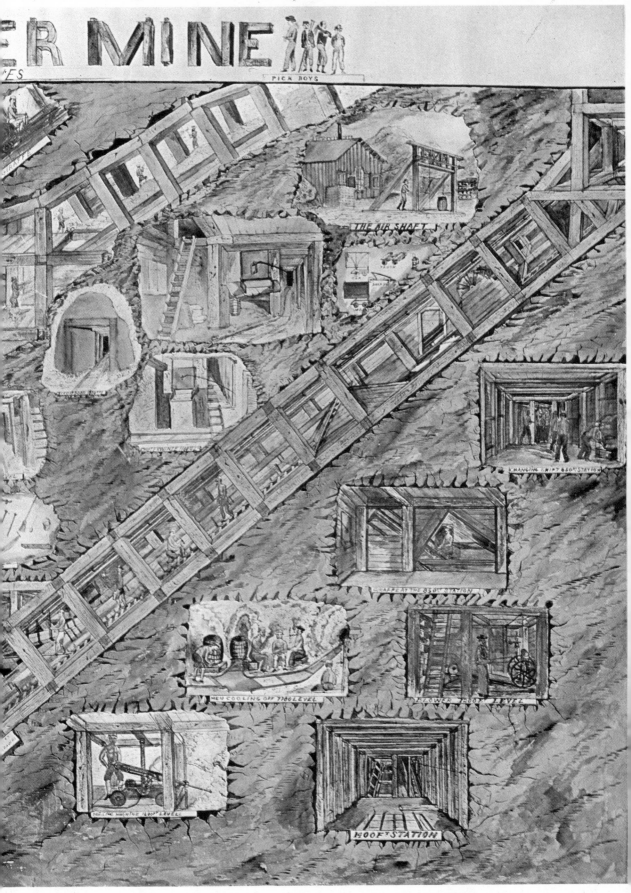

Opened in 1863, the Belcher Mine had produced $26,177,118 worth of precious ore by 1916.

souls when the miners were busy panning out the abundant gold. Today it contains fewer than several dozen persons. Among the buildings that remain are the Catholic church and a two-story hotel, notable mainly because they were built of lumber brought around Cape Horn, then by river boat to Sacramento, and finally by ox-cart to Hornitos.

The Western Pacific Railroad today winds its way through the bottom of the deep valley of the North Fork of the Feather River. Visitors today might not even notice an unimportant siding called Rich, but this was once Rich Bar, where several thousand miners toiled up and down the river and took out $3,000,000 while the gold lasted. There is no trace of the wood-and-canvas Empire Hotel, where Dame Shirley once lived, nor of the other makeshift buildings of that day, but the little cemetery still remains.

Some camps have entirely disappeared. You Bet, for example, was once a busy place, but its gold gave out and it became a ghost camp. Then the hydraulic miners found that great amounts of low-grade dirt still remained. The monitors started shooting their mighty streams, and You Bet vanished from the face of the earth, along with the hills on which it stood.

North San Juan is another former mining camp where a few people cling to their homes amid crumbling buildings on the edge of the deep canyon cut by the monitors. One old lady, asked about the days long ago, pointed to a spot somewhat above the scarred surface of the earth. "We used to live up yonder on the hill but it has all been piped away," she said. The expression "piped away" meant that hills and everything on them had washed away under the hammering of the monitors.

Miners have been gone from the old camps for more than a century, and the few remaining buildings of the ghost towns are crumbling into dust. Yet their disappearance brought about no financial loss, for California was destined to become a prosperous state even if gold had never been found there. Her farms yield far more wealth today than her gold mines ever have, while fisheries, forests, oil, and other minerals supply additional riches.

The lasting importance of the gold rush lay not in the millions of dollars in gold it produced, but in the fact that it happened when it did, at a time when America had just extended its borders to the Pacific Ocean. For of the thousands of forty-niners that the gold fever drew to California, many remained in the West, eventually helping to bring its vast lands securely into the Union as states.

Unknown to most of them, the California gold hunters were to achieve a more important place in American history as settlers rather than as miners. For the bearded forty-niner—with his red shirt and slouch hat, his trousers stuck into high boots, and his gold pan —played his greatest role not as a treasure seeker in El Dorado, but as a pioneer in the winning of the West.

Appendix

PICTURE CREDITS

The source of each picture used in this book is listed below, by page. When two or more pictures appear on one page, they are separated by semicolons. The following abbreviations are used:

CHS—California Historical Society
CSL—California State Library
CD—Carl Dentzel

HHL—Henry Huntington Library
NYHS—New York Historical Society
NYPL—New York Public Library

RH—Robert Honeyman
SCP—Society of California Pioneers
ZM—Zelda Mackay

Maps drawn expressly for this book by David Greenspan

Cover: "Placerville Mining" by Albertis Del Orient Browere—Knoedler Galleries. **Front End Sheet:** "Supply Wagon" by Eugene Camere—Pioneer Museum and Haggin Galleries, Stockton, Calif. **Half Title:** NYPL. **Title:** "The Lone Prospector" by Browere — Knoedler Galleries. **Contents:** Hist. Soc. of York County, Pa. **10-11** Wells Fargo Bank and Union Trust Co. **13** CHS. **14** (top) Yale University; (center) CSL; (bot. left) M. H. DeYoung Museum; (bot. right) CHS. **15** HHL. **16** (top) RHC; (bot.) CD. **17** University of California. **18** Wells Fargo Bank and Union Trust Co. **19** HHL. **20** Capitol, Sacramento. **23** Map drawn expressly for this book by David Greenspan. **24-5** (top left) CHS; (top right) RH; (bot. left) RH; (bot. right) CHS. **27** CHS. **28-9** NYPL. **31** CSL. **32-3** Museum of the City of New York. **34-5** Federal Reserve Bank of San Francisco. **36** (both) RH. **38** (top) RH; (bot.) Newark Museum. **40** ZM. **42-3** RH. **44-5** NYHS. **46-7** Hall Park McCullough. **48-9** Mrs. Angus Gordon Boggs. **50-1** (all) HHL. **53** Map by David Greenspan. **54** CD. **55** HHL. **56** (top) U. of Oregon Lib.; (center) Union Pacific R. R.; (bot.) NYPL. **59** Courtesy Second Bank-State Street Trust Co. **61** CHS. **62** ZM. **63** Dale Walden. **64-5** RH. **67** (top) ZM; (bot.) CSL. **69** (top) Knoedler Galleries; (bot.) Edmund B. Gerard. **70** Los Angeles County Museum. **72** John Rosekrans. **74-5** CD. **77** from *California Pictorial.* **78-9** CSL. **80** ZM. **81** Library of Congress. **82-3** RH. **84-5** CHS. **86-7** RH. **88** ZM. **89** ZM. **90** SCP. **91** CHS. **93** ZM. **95** Library of Congress. **96** Museum of the City of New York. **97** Museum of the City of New York. **98-9** (all) NYHS. **100** ZM. **102-3** CHS. **104** (top) NYHS; (bot.) Yale University Library. **105** NYPL. **106** (top) Stanford University; (bot.) NYHS. **107** (top) Gene Autry; (bot.) NYHS. **108** ZM. **110-11** "Sunday Morning in the Mines" by Charles Nahl—E. B. Crocker Art Gallery. **112** CHS. **114** RH. **115** RH. **117** CHS. **118-19** Library of Congress. **120** (both) Brown Bros. **121** NYPL. **122-23** "Preliminary Trial of a Horse Thief" by J. Mulvany—RH. **124-25** Smithsonian Institution. **126** (both) from *Mountains and Molehills* by F. Marryat—NYHS. **127** (top) NYPL; (bot.) SCP. **128** American Antiquarian Society. **130-31** Roger Lapham. **132** SCP. **133** NYHS. **135** (top) NYPL; (bot.) SCP. **137** ZM. **138** NYHS. **139** Nevada State Hist. Soc. **141** (top) Library of Congress; (bot.) CHS. **143** (both) by J. Lamsen — CHS. **144** Southern Pacific Co. **146-47** Mackay School of Mines, U. of Nevada. **149** CHS. **Back Cover:** (top to bot.) NYPL; Library of Congress; HHL; CHS; (lower right) NYHS.

BIBLIOGRAPHY

Bidwell, General John. *Echoes of the Past About California.* Chicago: Lakeside Press, 1928.

Bieber, Ralph P. "California Gold Mania," *Mississippi Valley Historical Review,* Vol. XXXV, No. 1 (June, 1948).

Borthwick, J. D. *Three Years in California.* London: William Blackwood & Sons, 1857.

Bruff, Joseph Goldsborough. *Gold Rush,* ed. Georgia Willis Read and Ruth Gaines. 2 Vols. New York: Columbia University Press, 1944.

Buckbee, Edna Bryan. *Pioneer Days of Angel's Camp.* Angel's Camp, Calif.: Calaveras Californian, 1932.

Canfield, Chauncey L. (ed.). *Diary of a Forty-niner.* Boston: Houghton Mifflin, 1920.

Caughey, John W. *Gold is the Cornerstone.* Berkeley: University of California, 1949.

—— (ed). *Rushing for Gold.* Berkeley: University of California, 1949.

Chalfant, W. A. *Gold, Guns and Ghost Towns.* Stanford, Calif.: Stanford University Press, 1947.

Clappe, Louise Amelia Knapp (Smith). *The Shirley Letters from the California Mines.* New York: Alfred A. Knopf, 1949.

Dana, Julian. *Sacramento: River of Gold.* New York: Holt, Rinehart & Winston, 1939.

——. *Sutter of California.* New York: Press of the Pioneers, 1934.

Gudde, Edwin G. (ed.). *Sutter's Own Story.* New York: G. P. Putnam's Sons, 1936.

Jackson, Joseph H. *Anybody's Gold.* New York: D. Appleton-Century, 1941.

—— (ed.). *Gold Rush Album.* New York: Charles Scribner's Sons, 1949.

Johnston, Philip. *Lost and Living Cities of the California Gold Rush.* Los Angeles: Touring Bureau of the Automobile Club of Southern California, 1948.

Leeper, David, *The Argonauts of Forty-nine.* Columbus, Ohio: Long's College Book Co., 1950.

Lewis, Oscar. *Sea Routes to the Gold Fields.* New York: Alfred A. Knopf, 1949.

McIlhany, Edward W. *Recollections of a 49'er.* Kansas City, Mo.: Hailman Printing Co., 1908.

McLeod, Alexander. *Pigtails and Gold Dust.* Caldwell, Idaho: Caxton Printers, Ltd., 1948.

Pierce, Hiram D. *A Forty-niner Speaks.* Oakland, Calif.: Keystone-Inglett, 1930.

Quaife, Milo (ed.). *Pictures of Gold Rush California.* Chicago: Lakeside Press, 1949.

Rickard, T. A. *A History of American Mining.* New York: McGraw-Hill, 1932.

Riesenberg, Felix, Jr. *The Golden Gate.* New York: Alfred A. Knopf, 1940.

Shinn, Charles Howard. *Mining Camps, A Study in American Frontier Government.* New York: Alfred A. Knopf, 1948.

Steele, Rev. John. *In Camp and Cabin.* Chicago: Lakeside Press, 1928.

Strong, Phil. *Gold in Them Hills.* Garden City, New York: Doubleday, 1957.

Sunset. *Gold Rush Country.* By the Editors of Sunset Books and Sunset Magazine. Menlo Park, Calif.: Lane Publishing Co., 1957.

White, Stewart E. *The Forty-niners.* New Haven: Yale University Press, 1920.

ACKNOWLEDGMENTS: The editors are deeply grateful to Mr. Archibald Hanna, Curator, Western Americana Collection, Yale University Library, for his unfailing assistance and guidance; Miss Jean Martin, California Historical Society, for her co-operation in furnishing material hitherto unpublished; and to Mr. Carl Dentzel, Southwest Museum, for generously making available his personal collection. In addition they wish expressly to thank the following individuals and organizations for their kindness in providing pictorial materials from their collections: Dr. Eliot Evans, Society of California Pioneers; Miss Irene Simpson, Wells Fargo Historical Bank; Mr. Allan Ottley, California State Library; Miss Elizabeth Clare, Knoedler Galleries; Mr. Robert Becker, Bancroft Library, University of California; and Mr. Robert Honeyman of New York.

Index

Bold face indicates pages on which illustrations appear